Daniel Ambort

The metalloendopeptidase meprin - degradomic approaches

Daniel Ambort

# The metalloendopeptidase meprin - degradomic approaches

## Principles and applications in protease substrate discovery

Südwestdeutscher Verlag für Hochschulschriften

**Impressum/Imprint (nur für Deutschland/ only for Germany)**
Bibliografische Information der Deutschen Nationalbibliothek: Die Deutsche Nationalbibliothek verzeichnet diese Publikation in der Deutschen Nationalbibliografie; detaillierte bibliografische Daten sind im Internet über http://dnb.d-nb.de abrufbar.
Alle in diesem Buch genannten Marken und Produktnamen unterliegen warenzeichen-, marken- oder patentrechtlichem Schutz bzw. sind Warenzeichen oder eingetragene Warenzeichen der jeweiligen Inhaber. Die Wiedergabe von Marken, Produktnamen, Gebrauchsnamen, Handelsnamen, Warenbezeichnungen u.s.w. in diesem Werk berechtigt auch ohne besondere Kennzeichnung nicht zu der Annahme, dass solche Namen im Sinne der Warenzeichen- und Markenschutzgesetzgebung als frei zu betrachten wären und daher von jedermann benutzt werden dürften.

Verlag: Südwestdeutscher Verlag für Hochschulschriften Aktiengesellschaft & Co. KG
Dudweiler Landstr. 99, 66123 Saarbrücken, Deutschland
Telefon +49 681 37 20 271-1, Telefax +49 681 37 20 271-0, Email: info@svh-verlag.de
Zugl.: Berne, University of Berne, Diss., 2007

Herstellung in Deutschland:
Schaltungsdienst Lange o.H.G., Berlin
Books on Demand GmbH, Norderstedt
Reha GmbH, Saarbrücken
Amazon Distribution GmbH, Leipzig
ISBN: 978-3-8381-0549-9

**Imprint (only for USA, GB)**
Bibliographic information published by the Deutsche Nationalbibliothek: The Deutsche Nationalbibliothek lists this publication in the Deutsche Nationalbibliografie; detailed bibliographic data are available in the Internet at http://dnb.d-nb.de.
Any brand names and product names mentioned in this book are subject to trademark, brand or patent protection and are trademarks or registered trademarks of their respective holders. The use of brand names, product names, common names, trade names, product descriptions etc. even without a particular marking in this works is in no way to be construed to mean that such names may be regarded as unrestricted in respect of trademark and brand protection legislation and could thus be used by anyone.

Publisher:
Südwestdeutscher Verlag für Hochschulschriften Aktiengesellschaft & Co. KG
Dudweiler Landstr. 99, 66123 Saarbrücken, Germany
Phone +49 681 37 20 271-1, Fax +49 681 37 20 271-0, Email: info@svh-verlag.de

Copyright © 2009 by the author and Südwestdeutscher Verlag für Hochschulschriften Aktiengesellschaft & Co. KG and licensors
All rights reserved. Saarbrücken 2009

Printed in the U.S.A.
Printed in the U.K. by (see last page)
ISBN: 978-3-8381-0549-9

*To my parents*
*Margrit and German Ambort*

# Contents

| | | |
|---|---|---|
| **Abbreviations** | | **IX** |

## Chapter 1

| | | |
|---|---|---|
| **1.** | **Introduction** | **1** |
| **1.1** | **Proteomics – classical and modern approaches** | **1** |
| 1.1.1 | Proteome and proteomics | 1 |
| 1.1.2 | Classical approaches | 1 |
| 1.1.3 | "Top-down" qualitative proteomics | 2 |
| 1.1.4 | "Bottom-up" qualitative proteomics | 3 |
| 1.1.5 | Quantitative proteomics | 3 |
| **1.2** | **Degradomics –protease substrate discovery** | **5** |
| 1.2.1 | Proteases – molecular switches in regulation of signalling ciruits | 5 |
| 1.2.2 | Degradomics – systems biology of the protease web | 5 |
| 1.2.3 | Degradomic techniques applied to different cell-based and cell-free systems – benefits and limitations | 6 |
| 1.2.4 | Degradomic studies point to novel functions for metalloendopeptidases | 7 |
| **1.3** | **Meprin – an astacin-like zinc-metalloendopeptidase** | **8** |
| 1.3.1 | The astacin family of metalloendopeptidases | 8 |
| 1.3.2 | Meprin – domain structure, biosynthesis and expression | 9 |
| 1.3.3 | Meprin – substrates, cleavage specificity and biological functions | 12 |
| **1.4** | **Aim of this project** | **13** |
| **1.5** | **References** | **15** |

# Chapter 2

## 2. Sample preparation of culture medium from Madin-Darby canine kidney cells — 22

| | | |
|---|---|---|
| 2.1 | Summary | 22 |
| 2.2 | Introduction | 23 |
| 2.3 | Materials | 25 |
| 2.3.1 | Cell culture | 25 |
| *2.3.1.1* | *Equipment* | *25* |
| *2.3.1.2* | *Solution and reagents* | *25* |
| 2.3.2 | Ultracentrifugation and ultrafiltration | 26 |
| *2.3.2.1* | *Equipment* | *26* |
| *2.3.2.2* | *Solutions and reagents* | *26* |
| 2.3.3 | Protein quantitation by BCA assay | 27 |
| *2.3.3.1* | *Equipment* | *27* |
| *2.3.3.2* | *Solutions and reagents* | *27* |
| 2.3.4 | Rehydration loading and isoelectric focusing | 27 |
| *2.3.4.1* | *Equipment* | *27* |
| *2.3.4.2* | *Solutions and reagents* | *28* |
| 2.4 | Methods | 29 |
| 2.4.1 | Cell culture | 29 |
| 2.4.2 | Ultracentrifugation and ultrafiltration | 30 |
| 2.4.3 | Protein quantitation by BCA assay | 31 |
| 2.4.4 | Rehydration loading and isoelectric focusing | 33 |
| 2.5 | Notes | 35 |
| 2.6 | References | 38 |

# Chapter 3

| | | |
|---|---|---|
| 3. | **Meprin – a metalloendopeptidase and its substrate repertoire – a degradomic approach** | **41** |
| 3.1 | **Summary** | **41** |
| 3.2 | **Introduction** | **42** |
| 3.3 | **Experimental procedures** | **44** |
| 3.3.1 | Cell culture and meprin activation by *in situ* trypsin treatment | 44 |
| 3.3.2 | Sample preparation of culture medium | 44 |
| 3.3.3 | One-dimensional gel electrophoresis | 45 |
| 3.3.4 | Two-dimensional gel electrophoresis | 45 |
| 3.3.5 | Staining and imaging | 46 |
| 3.3.6 | Image analysis | 46 |
| 3.3.7 | Protein identification by liquid chromatography- based tandem mass spectrometry and protein database searching | 47 |
| 3.3.8 | Functional classification | 49 |
| 3.3.9 | Immunoblotting | 49 |
| 3.4 | **Results** | **50** |
| 3.4.1 | Meprin activation by *in situ* trypsin treatment | 50 |
| 3.4.2 | Design and application of a novel two-dimensional gel image analysis strategy | 52 |
| 3.4.3 | Identification of dog protein orthologs by means of liquid chromatography- based tandem mass spectrometry, Phenyx-based and BLASTP-based protein database searching | 56 |
| 3.4.4 | Detection of neo N- and C-termini in peptide fragments | 64 |
| 3.4.5 | Functional clustering into biological process and molecular function | 65 |
| 3.4.6 | Validation of direct and indirect effects by immunoblotting follow up experiments | 66 |
| 3.5 | **Discussion** | **71** |
| 3.5.1 | Establishment of a novel two-dimensional gel-based degradomic approach | 71 |
| 3.5.2 | Key roles for human meprin in homeostasis of cell, cellular environment and in immune response | 72 |
| 3.6 | **References** | **75** |

# Chapter 4

| | | |
|---|---|---|
| **4.** | **The substrate degradome of meprin beta in human milk** | **80** |
| 4.1 | Summary | 80 |
| 4.2 | Introduction | 81 |
| 4.3 | Methods and materials | 83 |
| 4.3.1 | Human milk | 83 |
| 4.3.2 | *In vitro* digestion of human milk | 83 |
| 4.3.3 | *In vitro* digestion of chick tenascin-C | 83 |
| 4.3.4 | One-dimensional gel electrophoresis | 84 |
| 4.3.5 | Protein identification by liquid chromatography- based tandem mass spectrometry and protein database searching | 84 |
| 4.3.6 | Functional classification | 85 |
| 4.3.7 | Immunoblotting | 85 |
| **4.4** | **Results** | **86** |
| 4.4.1 | *In vitro* digestion of human milk by recombinant human meprin beta | 86 |
| 4.4.2 | Functional classification of human milk proteins identified by liquid chromatography-based tandem mass spectrometry and protein database searching | 88 |
| 4.4.3 | *In vitro* cleavage of purified chick tenascin-C splicing variants by recombinant human meprin alpha and beta | 90 |
| 4.4.4 | Immunoblotting of recombinant human meprin alpha- and beta-generated chick tenascin-C digests with domain-specific monoclonal antibodies | 93 |
| **4.5** | **Discussion** | **97** |
| 4.5.1 | Meprin beta and its substrate repertoire in human milk | 97 |
| 4.5.2 | Human tenascin-C a novel component in human milk | 98 |
| 4.5.3 | Distinct functions for the two meprin subunits alpha and beta exemplified by tenascin-C proteolytic processing | 98 |
| **4.6** | **References** | **101** |

| 5. | **Outlook** | **104** |
|---|---|---|
| 6. | **Acknowledgements** | **106** |

# Abbreviations

| | |
|---|---|
| ADAM | a disintegrin and metalloprotease |
| BMP-1 | bone morphogenetic protein 1 |
| CRD | carbohydrate recognition domain |
| 2-D | two-dimensional |
| 2-DE | two-dimensional gel electrophoresis |
| DIGE | differential in-gel electrophoresis |
| ECM | extracellular matrix |
| EGF | epidermal growth factor |
| ESI | electrospray ionization |
| FN-III | fibronectin type III |
| FWHM | full width at half maximum |
| hmeprin | human meprin |
| ICAT | isotope-coded affinity tag |
| IEF | isoelectric focusing |
| IPG | immobilized pH gradient |
| LC-MS/MS | liquid chromatography-based tandem mass spectrometry |
| MDCK | Madin-Darby canine kidney |
| MMP | matrix metalloproteinase |
| $M_r$ | molecular weight |
| MS | mass spectrometry |
| mTLD | mammalian tolloid |
| MudPIT | multidimensional protein identification technique |
| PABA-peptide | N-benzoyl-L-tyrosyl-p-aminobenzoic acid |
| p$I$ | isoelectric point |
| PKC | protein kinase C |
| PMA | phorbol 12-myristate 13-acetate |
| rhmeprin | recombinant human meprin |
| RP | reversed-phase |
| SDS-PAGE | sodium dodecyl sulfate-polyacrylamide gel electrophoresis |
| S/N | signal-to-noise ratio |
| TACE | tumor necrosis factor-α converting enzyme |
| TN-C | tenascin-C |

# Chapter 1

## 1. Introduction

### 1.1 Proteomics – classical and modern approaches

#### 1.1.1 Proteome and proteomics

A proteome represents an expressed set of proteins encoded by the genome (1). Hence a proteome consists of several thousand proteins. However, not all proteins are expressed at the same time, and in a living cell changes are happening continuously. This makes it possible to create a proteome map like a static gene map. Therefore during an experiment, a series of samples is analyzed, and the quantitative changes of expression levels are monitored. This approach is called proteome analysis or proteomics. The term proteomics was coined in 1995 and was defined as the large-scale characterization of the entire protein complement of a cell line, tissue or organism (2-4). Many different areas are grouped under the rubric of proteomics including functional genomics, post-translational modification studies, protein expression profiling, protein-protein interaction studies, proteome mining and structural proteomics (5).

#### 1.1.2 Classical approaches

Traditionally, two-dimensional gel electrophoresis (2-DE), introduced by O'Farrell and Klose in 1975 (6, 7), enabled separation of complex protein mixtures into individual protein species according to their net charge (p$I$) in the first dimension by isoelectric focusing (IEF) and according to their molecular weight ($M_r$) in the second dimension by sodium dodecyl sulfate-polyacrylamide gel electrophoresis (SDS-PAGE) (8). In the conventional approach, IEF was performed in carrier ampholyte-generated pH gradients, which moved towards the cathode upon prolonged focusing time. This "cathodic drift" phenomenon was thereafter remedied by non-equilibrium pH gradient gel electrophoresis and finally eliminated with the invention of fixed immobilized pH gradients (IPG) (9-13). The development of microanalytical techniques, namely, Edman sequencing (14-16) and mass spectrometry (MS) (17-19) enabled identification of proteins at amounts available from a single 2-D gel. The

Edman sequencing technique allowed the unambiguous identification of proteins in 2-DE after blotting onto membranes. Edman degradation takes place in a sequencer that automatically degrades proteins and peptides from their N-terminal end by a stepwise chemical procedure using phenylisothiocyanate (20). The advances of MS instruments allowed the identification of proteins from 2-D gels by "peptide mass fingerprinting" of a tryptic protein digest, namely by comparison of the measured masses of the individual peptides in the mixture with the theoretically derived peptide masses calculated from all proteins in the protein database (Swiss-Prot, NCBI) (17-19). The peptide mixture derived was analyzed by matrix-assisted laser desorption ionization-MS or electrospray ionization (ESI)-MS (21, 22).

### *1.1.3 "Top-down" qualitative proteomics*

2-DE advanced to the core technology of proteome analysis (12, 13) and was brought from art to craft in an industrial standard. In 2-DE post-translational modifications can be studied by a protein expression profiling approach. A number of post-translational modifications can occur in a gene product, like truncation, phosphorylation, different kinds of glycosylation, acetylation, etc. and will alter the separation parameters, the net charge and the mass. 2-DE has a number of important features: extremely high resolving power, tolerance to crude protein mixtures and tolerance to relatively high sample loads. In addition, 2D gels are efficient fraction collectors and proteins are protected inside the gel matrix for analysis (23). Despite these advantageous features 2D gels suffer from some intrinsic limitations. A major problem of 2-DE are co-migrating proteins in the same spot (24, 25). However, the use of zoom or narrow pH range gels will increase resolution and decrease the probability of protein co-migration (26). Another problem is the dynamic range of protein expression in cells and tissues (27). The most abundant proteins will then prevent resolution of low copy number proteins such as transcription factors, protein kinases and other regulatory proteins. Therefore enrichment and prefractionation strategies are needed to reach the less abundant proteins (28). A third problem is linked with the protein extraction and solubility during 2-DE (29). Poorly water-soluble proteins such as transmembrane proteins resist entry into an IPG strip. Last but not least, very basic proteins ($pI$ >11) and high molecular weight proteins (>200 kDa) are not easy to analyze by 2-DE (23).

## 1.1.4 "Bottom-up" qualitative proteomics

An alternative 2-D-free approach to identify soluble and membrane proteins in mixtures involves the use of liquid chromatography-tandem MS (LC-MS/MS). In this strategy, proteins are proteolytically digested before they are separated (30, 31). The tandem mass spectra of peptides can then be used to search sequence databases to identify proteins by matching amino-acid sequences to each spectrum (30). Tandem mass spectrometers can select peptide ions for the process of "collision-induced dissociation", which can be used to provide information about the peptide sequence. Under computer control, the process is both fast and efficient, but not selective for specific peptide ions because any ion of sufficient abundance is selected for analysis. This situation results in a semi-random sampling process when complex mixtures are analyzed (32). The proteolysis of protein mixtures creates complex mixtures of peptides. Due to the complexity of these protein mixtures fractionation is achieved by integrating a two-dimensional liquid chromatography separation technique with a tandem mass spectrometer to create a multidimensional protein identification technique (MudPIT), and this was subsequently named shotgun proteomics (33, 34). The method produces a serial separation on the basis of charge interactions with a strong cation-exchange resin followed by hydrophobic interactions with a reversed-phase (RP) support. Peptides are eluted from the cation-exchange resin using increasing salt concentrations, and they are subsequently trapped on the RP support. The salt is washed from the system and then the peptides are eluted into a tandem mass spectrometer using an increasing gradient of non-polar organic solvent. The MudPIT-derived proteome is a raw list of proteins present in the sample, without any accurate quantitative aspect.

## 1.1.5 Quantitative proteomics

Nowadays there exists a trend towards quantitative proteomics to study protein levels between samples that differ by some variable. The two most prominent techniques are differential in-gel electrophoresis (DIGE) and isotope-coded affinity tag (ICAT) labelling (35, 36). DIGE is a technique that labels complex protein mixtures with different fluorescent tags prior to conventional 2-DE. Up to three different samples can be labelled and mixed together and then separated on a single 2-D gel. Cyanine dyes are used to label the protein from different samples with dyes of different excitation and emission wavelengths (35). The labelling method used is lysine labelling which labels the proteins via the epsilon amino group of lysine. The dyes are matched for charge and molecular weight ensuring that the same protein found in each sample will migrate to the same position on a 2-D gel. One advantage of this method over conventional methods is that since the samples are exposed to the same chemical environments and electrophoretic conditions, co-migration is guaranteed for identical proteins from the separate samples and analysis of sample differences is therefore simplified. The ratio of protein expression is always obtained from a sinlge gel and an internal standard can be used in each gel significantly reducing gel to gel variation of protein ratio measurements. Matching proteins between gels allows ratio measurements to be compensated from a number of different samples. A specific software package (DeCyder) has been developed to exploit the unique advantages of this technology allowing an automated approach to the analysis of differences found within and between gels. In the ICAT strategy two protein mixtures representing two different cell states are treated with isotopically light and heavy ICAT reagents, respectively. The ICAT reagent consists of three elements: an affinity tag (biotin), which is used to isolate ICAT-labelled peptides; a linker that can incorporate stable isotopes; and a reactive group with specificity towards thiol groups (cysteines). The reagent exists in two forms, heavy (contains eight deuteriums) and light (contains no deuteriums) (36). An ICAT reagent is covalently attached to each cysteinyl residue in every protein. The protein mixtures are combined and proteolytically cleaved to peptides, and ICAT-labelled peptides are isolated utilizing the biotin tag. These peptides are separated by liquid chromatography. A pair of ICAT-labelled peptides is chemically identical and is easily visualized because these peptides essentially co-elute, and there is an 8 Da mass difference measured in a scanning mass spectrometer. The ratios of the original amounts of proteins from the two different cell states are strictly maintained in the peptide fragments. The relative quantification is determined by the ratio of the peptide pairs. Every other scan is devoted to fragmenting and then recording sequence information about an eluting peptide (tandem mass spectrum). The protein is identified by computer-searching the recorded sequence information against large protein databases.

## 1.2 Degradomics – protease substrate discovery

### 1.2.1 Proteases – molecular switches in regulation of signalling circuits

The key feature of any protease is its ability to hydrolyze peptide bonds in target protein substrates. Cleavage of target peptide bonds upon protease attack is an irreversible process leading to degradation products with functions distinct from their parental forms. Thus complete degradation of substrates potentially cleaved at any scissile bond available would have deleterious effects on the cell and its extracellular environment, namely, the extracellular matrix (ECM) surrounding cells. Proteolysis of substrates is termed processing when the specific and efficient cleavage of target proteins occurs only at one or two sites. For example, it has long been recognized that processing is crucial to the selectivity of the clotting cascade and in trypsinogen activation (37-39). Moreover, the proteolytic shedding of ectodomains has emerged as crucial regulator of growth factors and cytokines such as transforming growth factor-$\alpha$ or tumour necrosis factor-$\alpha$ (40-42). In addition, processing of chemokines such as monocyte chemoattractant protein-3 regulated leukocyte trafficking and inflammatory responses (43, 44). Another important proteolytic event is the exposure of cryptic epitopes such as in laminin-5 following matrix metalloproteinase (MMP) cleavage to enhance cell migration (45). Last but not least, upon MMP cleavage the release of cryptic neoproteins such as angiostatin from plasminogen and endostatin from collagen type XVIII, respectively, demonstrated novel roles for these peptides in inhibition of angiogenesis (46, 47). In conclusion, by precisely controlling cell function, proteases are essential components of signalling pathways and thus act as molecular switches in regulation of signalling circuits at the cell surface and in the extracellular milieu (48).

### 1.2.2 Degradomics – systems biology of the protease web

In the past protease substrate discovery was rather haphazard and susceptibility to proteolysis was tested by *in vitro* cleavage assays using singly selected protein targets. In respect of the complexity of proteases encoded by the human genome, namely, 21 aspartate-, 148 cysteine-, 194 metallo-, 175 serine- and 28 threonine proteases, encompassing 2% of the genes, these traditional substrate finding concepts may not reflect the *in vivo* situation (48). Moreover, the highly interdependent nature of proteolytic systems *in vivo* may represent a protease web similar to the world-wide web connecting diverse proteolytic pathways and proteases of different families that can form information conduits determining the functional state of the protease web (49). Thus proteolytic activity towards a particular substrate in a cell or tissue corresponds to the sum of net activities of all proteases present that directly cleave the substrate or indirectly modify the activity of the responsible proteases. Therefore protease substrate discovery may be applied to cell-based systems in order to study proteolytic activity on a system-wide scale and under physiologic conditions. Recently, the term degradome was coined to define the protease and protease substrate repertoires of a cell, tissue or organism at a specific time on a proteome-wide level (50). The term degradome was defined as the large-scale identification of any protease expressed by the system studied. The degradome of a protease may then constitute its substrate repertoire. Hence, degradomics may encompass all genomic and proteomic investigations and techniques regarding the genetic, structural and functional identification and characterization of proteases, and their substrates and inhibitors, that are present in an organism.

## 1.2.3 Degradomic techniques applied to different cell-based and cell-free systems – benefits and limitations

Upon the introduction of a plethora of techniques and applications emerged to study the degradome by means of degradomics. For example, LC-MS/MS in combination with ICAT labelling or 2-DE and matrix-assisted laser desorption/ionization time-of-flight-MS were applied to identify the substrate degradome of MMP-14 (51, 52). In the ICAT approach MMP-14 or inactive E240A mutant transfected MDA-MB-231 breast carcinoma cells were used to quantitatively find shed or degraded proteins in the extracellular milieu or culture medium. Although successfully applied, this method may not discern post-translational modifications of identical protein species and targets only cysteine-labelled peptides. In addition, some conceptual problems may arise: ICAT-based approaches compare pairs of peptides and therefore it is not possible to discover cleaved protein fragments with neo N- or C-termini. The second group used human plasma as a polysubstrate to be treated with recombinant MMP-14 catalytic domain. Using human plasma as a biologically relevant substrate may be questionable due to the broad dynamic range in abundance of different protein classes (53). In another targeted proteomic approach bone marrow-derived monocytic cells were stimulated in the presence or absence of a pan-metalloprotease inhibitor by phorbol 12-myristate 13-acetate (PMA) to induce cell surface protein ectodomain shedding. Conditioned media were lectin-affinity purified and deglycoslytated prior to separation with

2-DE and protein identification by LC-MS/MS (54). This pre-fractionation step allowed for concentration and selection of glycoproteins shed by membrane-bound tumor necrosis factor-α converting enzyme (TACE/ADAM-17). Very recently, this method was upgraded to specifically label conditioned media of a disintegrin and metalloprotease (ADAM)-17 transfected A431 human epithelial carcinoma cell line either treated with the hydroxamic acid-derived non-specific metalloprotease inhibitor batimastat or non-treated. Lectin-affinity purification, multiple labelling of media with different cyanine dyes and sample pooling enabled the separation with DIGE within the same run (55). DIGE allowed for quantitative and differential display of glycoproteins shed by membrane-bound ADAM-17. Both approaches were very efficient in identifying shed membrane proteins present at very low abundant levels in cells. Nevertheless lectin-affinity purification with wheat germ agglutinin selects for oligosaccharides containing sialic acid or terminal N-acetylglucosamine. Hence, N-glycosylated, O-glycosylated or non-glycosylated proteins and degradation products not bearing those ligands may escape from such a pre-fractionation procedure. Interestingly, in none of these reports were novel potential cleavage substrates systematically grouped into specific, functional categories in order to draw meaningful conclusions on a whole system's response upon its perturbation. Thus functional clustering may remedy misleading data interpretation. Moreover, applying pre-fractionation techniques, namely, selection for cysteine-containing peptides or glycoproteins, may allow for higher resolution capacity but at the cost of information loss.

## *1.2.4 Degradomic studies point to novel functions for metalloendopeptidases*

A variety of hitherto unkown substrates for the metzincin metalloendopeptidases ADAM-17 and MMP-14 were found in conditioned media using different cell-based systems (51, 54, 55). Alternatively, a complex protein mixture, human plasma, was digested with recombinant MMP-14 in a cell-free system (52). A plethora of novel substrates were described that were shedd off the membrane by ADAM-17, namely, the cell-surface receptor LDLr, the transmembrane glycoprotein SHPS-1, Jagged 1 a ligand for the receptor Notch 1, the transferrin receptor, the immunoglobulin superfamily member leukocyte cell adhesion molecule and the desmosomal cadherin desmoglein 2 (54, 55). In the cell-based degradomic approach using the membrane-bound form of MMP-14 the following novel substrates were reported: neutrophil chemokine interleukin-8, secretory leukocyte protease inhibitor, the cytokine tumour necrosis factor-α, death receptor-6 and connective tissue growth factor (51).

Furthermore, digestion of human plasma with the recombinant MMP-14 catalytic domain demonstrated processing of previously unkown substrates such as apolipoprotein A-I, apolipoprotein E and plasma gelsolin (52). From these degradomic studies it became obvious that metalloendopeptidases are key modulators of diverse signalling pathways and not ECM destroying bulldozers (56). For example, the major roles for the MMP family are the control of cellular responses critical to homeostatic regulation of the extracellular environment and the immune response (57, 58).

## 1.3 Meprin – an astacin-like zinc-metalloendopeptidase

### *1.3.1 The astacin family of metalloendopeptidases*

Six mechanistic classes of proteases have been described in living organisms: serine, threonine, cysteine, aspartic, glutamic and metalloendopeptidases. The metalloendopeptidases are hydrolases in which the nucleophilic attack on a peptide bond is mediated by a water molecule that is activated by a divalent metal cation in the active site, usually a zinc, but sometimes cobalt or manganese. One family of zinc-metalloendopeptidases that emerged in the 1990s is the astacin family named after the prototype astacin an endopeptidase isolated from the digestive tract of the freshwater crayfish *Astacus fluviatilis* (59-62). Astacins are characterized by an elongated zinc binding motif in their protease domain with the consensus sequence HEXXHXXGXXH. The X-ray crystal structure of astacin has been solved to 1.8 Å resolution revealing the catalytic zinc being located at the bottom of a long deep active site cleft ligated by three histidine residues, a tyrosine residue and a water molecule polarized by a glutamic acid residue (63-65). Members of the astacin family are found throughout the animal kingdom, in bacteria and also in humans and have been implicated in diverse developmental and maturation processes (61, 66). Most of these proteases have a multidomain structure including an astacin-like catalytic domain, one or more copies of an epidermal growth factor (EGF)-like domain and/or a CUB (complement subcomponents C1r/C1s, embryonic sea urchin protein Uegf, bone morphogenetic protein 1) domain. CUB and EGF-like domains are non-catalytic and thought to be involved in protein-protein or protein-substrate interactions. Astacin the smallest member of the family contains no additional C-terminal domain and is a soluble digestive enzyme in crustaceans. Several members of the family are implicated in hatching processes of embryos such as the teleostean enzymes choriolysin L and H (LCE, HCE1 and 2) which degrade the egg shell/chorion of fish embryos (67, 68). HMP1 an astacin-like protease of the polyp *Hydra vulgaris* plays a role in morphogenetic processes such as hydra head regeneration and degradation of extracellular matrix proteins in the developing tentacles (69, 70). Bone morphogenetic protein 1 (BMP-1) has been shown to be implicated in cartilage and bone formation in vertebrates by activating latent growth factors of the transforming growth factor-β superfamily (71). BMP-1 is identical with the procollagen C protease an enzyme responsible for processing of procollagen precursors of the major fibrillar collagens (type I, II and III) (72). The counterpart of mammalian BMP-1 and the transforming growth factor-βs in *Drosophila* named *tolloid* and *decapentaplegic*, respectively, are intimately involved in dorsal-ventral patterning of embryos (73). Meprins isolated from rodent kidney and human intestinal brush border preparations are original members of the astacin family whose cDNA were cloned and sequenced (74-79).

## *1.3.2 Meprin – domain structure, biosynthesis and expression*

Human meprin (hmeprin) is an astacin-like metalloendopeptidase first discovered in the early 1980s due to its ability to hydrolyze the substrate N-benzoyl-L-tyrosyl-p-aminobenzoic acid (PABA-peptide) (80). At the same time the PABA-peptide was used as a substrate in clinical diagnosis of exocrine pancreatic insufficiency (81, 82). These clinical tests were based on the activity of chymotrypsin on the PABA-peptide. Interestingly, residual non-pancreatic hydrolytic activity towards the PABA-peptide was detected in patients after total pancreatectomy (removal of the pancreas) (83). This led to the discovery of a PABA-peptide hydrolyzing enzyme in brush border membranes of human small intestinal epithelial cells. At the same time a similar metalloendopeptidase from mouse and rat intestinal tissue was purified and named meprin (metal endopeptidase from renal tissue) (84-86). Cloning and sequencing of the hmeprin cDNA revealed an astacin-like metalloendopeptidase with a multidomain protein structure (79, 87). Two similar subunits were identified termed meprinα and meprinβ with molecular masses of 95 and 105 kDa, respectively (88, 89). The overall domain structure of meprin comprises a N-terminal pro-peptide that has to be cleaved off for enzyme activation, the astacin-like protease domain, followed by a MAM (meprin/ A5 protein/ receptor tyrosine phosphatase μ) domain and a TRAF (tumor necrosis factor receptor-associated factor) domain. C-terminally of the MAM and TRAF domains which are thought to be protein interaction domains is an EGF-like domain, a transmembrane domain and a cytosolic region (Fig. 1). The C-terminal cytoplasmic tail of meprinβ may be phosphorylated at serine position 687 by protein kinase C (PKC) upon PMA treatment (90).

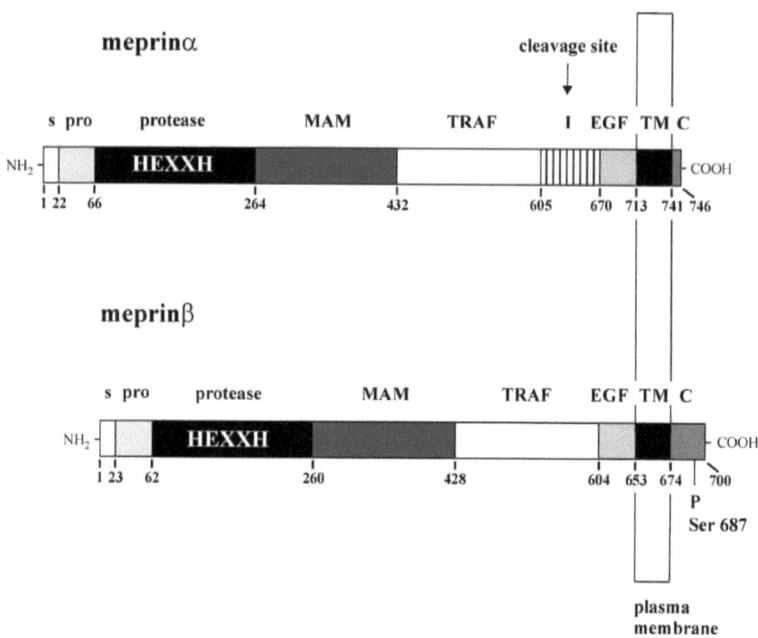

Fig.1. Protein domain organization of meprin alpha and beta deduced from the human cDNA sequence. Numbers indicate the position of individual domains depicted as follows: s, signal peptide; pro, propeptide; protease, astacin-like protease domain (HEXXH: zinc-metalloprotease consensus sequence); MAM, meprin/ A5 protein/ receptor tyrosine phosphatase µ domain; TRAF, tumor necrosis factor receptor-associated factor domain; I, inserted domain; EGF, epidermal growth factor-like domain; TM, transmembrane domain; C, cytoplasmic domain. Meprinα is processed intracellularly in the specific I-domain as indicated by an arrow. The C-terminal cytoplasmic tail of meprinβ can be phosphorylated at serine position 687.

The cDNA of human and rodent meprin subunits has been transfected and expressed in different cell systems (79, 87, 89, 91, 92). Biosynthesis studies in transfected COS-1 cells as well as metabolic labelling of organ cultures of human small intestinal mucosa showed that meprin subunits associate by interchain disulfide bonds to homo- or heterodimeric proteins in the endoplasmatic reticulum (ER) (89, 93). Expression of the α-subunit in MDCK (Madin-Darby canine kidney) cells revealed that the transmembrane anchor and cytosolic region of meprinα is intracellularly processed leading to the secretion of meprinα zymogens into the culture medium (Fig. 2) (87).

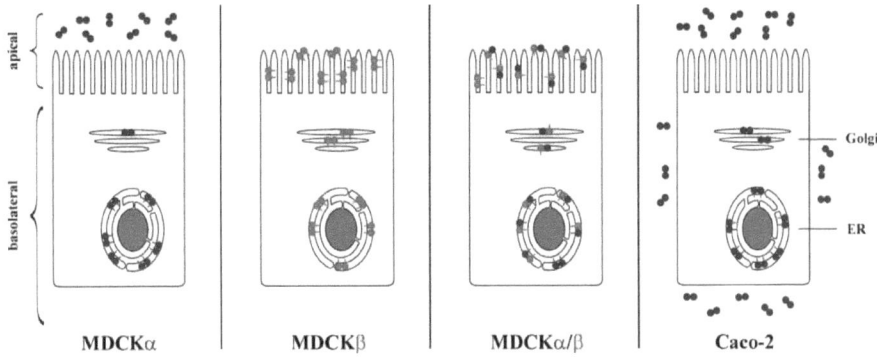

**Fig. 2. Biosynthesis of meprin alpha and beta subunits in polarized epithelial cells.**
MDCK cells were stably transfected with full length cDNAs of human meprinα (MDCKα), meprinβ (MDCKβ), or both meprinα and meprinβ (MDCKα/β) and analyzed for biosynthesis of enzyme subunits by metabolic labelling of cells and media. In MDCKα cells the α-subunit forms homodimers and is processed intracellularly, thereby loosing its membrane anchor. The zymogen thereafter is secreted apically. In MDCKβ cells homodimers of meprinβ are transported through ER and Golgi apparatus and get inserted into the apical plasma membrane. In MDCKα/β cells meprinα- and β-subunits form homodimeric and heterodimeric proteins which are transported to the apical plasma membrane. Meprinα is retained at the cell surface via association with transmembrane meprinβ. In contrast, differentiated, polarized Caco-2 cells endogenously express meprinα that is secreted in a non-polarized fashion from apical as well as from basolateral sides.

Essential for the cleavage is the inserted domain (I-domain) which is present only in the α-subunit (94). In contrast, expression of the β-subunit in MDCK cells results in the formation of homodimers that are readily transported and bound to the plasma membrane (79). The expression of both subunits in these cells leads to the assembly of α- and β-heteromers at the cell surface. Thus hmeprinα exists as secreted form or is bound to the β-subunit at the cell surface. In contrast, meprinβ is predominantly localized at the plasma membrane only a minor part is secreted into the culture medium (79). In the human gut both subunits are expressed in small intestinal epithelial cells. Here, they are localized at the apical brush border membrane, the α-subunit being bound the to β-subunit. In the colon only meprinα but not meprinβ was found to be expressed by colonocytes (95). Hence in colon meprinα is secreted entirely into the colon lumen. Beside the expression of meprin in normal intestinal epithelial cells meprin also seems to play an important role under pathological conditions. Hmeprin mRNA has been detected in leukocytes of the intestinal lamina propria suggesting a possible role in inflammatory processes of the gut (95). In addition, an isoform of the meprinβ mRNA comprising the identical protein coding region was found to be expressed in a variety of cancer cells, namely, human breast cancer cell lines (MCF-7 and SK-BR-3), the human osteosarcoma cell line U2 Os and the human pancreatic cancer cell line BxPC-3 (96). Moreover, an altered localization of meprinα has been proposed to promote acute renal failure in inbred mice (97). Non-polarized secretion of hmeprinα from colorectal cancer cells (Caco-2) has been shown to result in an accumulation of meprinα in the stroma of colorectal cancers (Fig. 2) (98).

## *1.3.3 Meprin – substrates, cleavage specificity and biological functions*

A first step towards the elucidation of the biological function of meprin was achieved with testing of putatively cleavable polypeptide substrates. A variety of protein and peptide substrates were processed *in vitro*; biologically active peptides such as bradykinin, angiotensins I and II (88), polypeptide hormones, insulin B-chain and parathyroid hormone (99), as well as gastrointestinal peptides, namely, gastrin-releasing peptide fragment 14-27 and gastrin 17 (100), ECM components, collagen type IV, fibronectin and laminin-nidogen (101, 102), cytokines such as osteopontin (100), interleukin-1β and the chemokine monocyte chemoattractant protein-1 (103, 104). This suggests that meprin may be involved in the clearance of vasoactive peptides and polypeptide hormones from blood plasma, in the regulation of cell movement, the secretory activity and growth of intestinal tract and pancreas, in tissue remodelling processes, and finally, in the innate immune response. In addition, marked differences between α- and β-subunits in substrate and peptide bond specificity were reported and may point to distinct functions for the two forms (100). Meprinα selects for small (e.g. serine, alanine, threonine) or hydrophobic (e.g. phenylalanine) residues in P1 and P1' sites, and proline in P2' position. Meprinβ prefers acidic amino acids in the P1 and P1' sites and selects against basic residues at P2' and P3' (100). In conclusion, protease-substrate discovery executed by these *in vitro* cleavage assays was rather haphazard. Thus meprin and its substrate repertoire may be studied in a complex biological context to identify physiologically relevant substrates.

## 1.4 Aim of this project

In recent years the introduction and application of degradomic techniques led to the discovery that metalloendopeptidases are molecular switches in signalling circuits and not extracellular matrix destroying bulldozers. In the past, protease-substrate discovery was rather haphazard and executed by *in vitro* cleavage assays using singly selected targets. Nowadays, with the rapidly growing field of proteomics, a protease and its substrate repertoire may be studied in complex biological systems; integrating information of a whole system's response to a protease's action. Thus the substrate degradome of meprin may be studied in a complex biological context to identify physiologically relevant substrates. This goal may be achieved by using different cell-based and cell-free systems to find novel hitherto unknown substrates that may elucidate key roles for meprin in various biological processes. The work described herein is subdivided into three parts:

### I. Sample preparation of culture medium from Madin-Darby canine kidney cells

A reproducible, standardized and simple sample preparation methodology is the key to successful 2-DE. This chapter describes step-by-step the sample preparation of culture medium from MDCK cells. Tips and tricks are given to circumvent possible pitfalls (Chapter 2).

### II. Meprin – a metalloendopeptidase and its substrate repertoire – a degradomic approach

Here, we report the first proteomic approach applied to meprin, an astacin-like metalloendopeptidase, to determine physiologic substrates in a whole cell system. Hmeprin$\alpha/\beta$ expressed on the cell surface of MDCK cells was activated by limited trypsin treatment. Culture media of these cells were subjected to 2-DE, image analysis and LC-MS/MS to identify proteins cleaved by meprin (Chapter 3).

### III. The substrate degradome of meprin beta in human milk

Here, we report a degradomic approach to determine the substrate repertoire of the astacin-like metalloendopeptidase meprin beta in human milk by combining *in vitro* cleavage assays, one-dimensional gel electrophoresis and LC-MS/MS for subsequent protein identification (Chapter 4).

## 1.5 References

1. Wilkins MR, Pasquali C, Appel RD, Ou K, Golaz O, Sanchez JC, Yan JX, Gooley AA, Hughes G, Humphery-Smith I, Williams KL, Hochstrasser DF. From proteins to proteomes: large scale protein identification by two-dimensional electrophoresis and amino acid analysis. *Biotechnology* 1996;14:61-5.
2. Wasinger VC, Cordwell SJ, Cerpa-Poljak A, Yan JX, Gooley AA, Wilkins MR, Duncan MW, Harris R, Williams KL, Humphery-Smith I. Progress with gene-product mapping of the Mollicutes: Mycoplasma genitalium. *Electrophoresis* 1995;16:1090-4.
3. Wilkins MR, Sanchez JC, Gooley AA, Appel RD, Humphery-Smith I, Hochstrasser DF, Williams KL. Progress with proteome projects: why all proteins expressed by a genome should be identified and how to do it. *Biotechnol Genet Eng Rev* 1996;13:19-50.
4. Anderson NG, Anderson NL. Twenty years of two-dimensional electrophoresis: past, present and future. *Electrophoresis* 1996;17:443-53.
5. Graves PR, Haystead TA. Molecular biologist's guide to proteomics. *Microbiol Mol Biol Rev* 2002;66:39-63.
6. O'Farrell PH. High resolution two-dimensional electrophoresis of proteins. *J Biol Chem* 1975;250:4007-21.
7. Klose J. Protein mapping by combined isoelectric focusing and electrophoresis in mouse tissues. A novel approach to testing for induced point mutations in mammals. *Humangenetik* 1975;26:231-43.
8. Lämmli UK. Cleavage of structural proteins during the assembly of the head of bacteriophage T4. *Nature* 1970;227:680-5.
9. O'Farrell PZ, Goodman HM, O'Farrell PH. High resolution two-dimensional electrophoresis of basic as well as acidic proteins. *Cell* 1977;12:1133-42.
10. Bjellqvist B, Ek K, Righetti PG, Gianazza E, Görg A, Westermeier R, Postel W. Isoelectric focusing in immobilized pH gradients: principle, methodology and some applications. *J Biochem Biophys Methods* 1982;6:317-39.
11. Görg A, Postel W, Günther S. The current state of two-dimensional electrophoresis with immobilized pH gradients. *Electrophoresis* 1988;9:531-46.
12. Görg A, Obermaier C, Boguth G, Harder A, Scheibe B, Wildgruber R, Weiss W. The current state of two-dimensional electrophoresis with immobilized pH gradients. *Electrophoresis* 2000;21:1037-53.
13. Görg A, Weiss W, Dunn MJ. Current two-dimensional electrophoresis technology for proteomics. *Proteomics* 2004;4:3665-85.
14. Matsudaira P. Sequence from picomole quantities of proteins electroblotted onto polyvinylidene difluoride membranes. *J Biol Chem* 1987;262:10035-8.
15. Aebersold RH, Leavitt J, Saavedra RA, Hood LE, Kent SB. Internal amino acid sequence analysis of proteins separated by one- or two-dimensional gel electrophoresis after in situ protease digestion on nitrocellulose. *Proc Natl Acad Sci* 1987;84:6970-4.
16. Rosenfeld J, Capdevielle J, Guillemot JC, Ferrara P. In-gel digestion of proteins for internal sequence analysis after one- or two-dimensional gel electrophoresis. *Anal Biochem* 1992;203:173-9.
17. Yates 3rd JR, Speicher S, Griffin PR, Hunkapiller T. Peptide mass maps: a highly informative approach to protein identification. *Anal Biochem* 1993;214:397-408.

18. James P, Quadroni M, Carafoli E, Gonnet G. Protein identification in DNA databases by peptide mass fingerprinting. *Protein Sci* 1994;3:1347-50.
19. Cottrell JS. Protein identification by peptide mass fingerprinting. *Pept Res* 1994;7:115-24.
20. Edman P, Begg G. A protein sequenator. *Eur J Biochem* 1967;1:80-91.
21. Karas M, Hillenkamp F. Laser desorption ionization of proteins with molecular masses exceeding 10,000 daltons. *Anal Chem* 1988;60:2299-301.
22. Fenn JB, Mann M, Meng CK, Wong SF, Whitehouse CM. Electrospray ionization for mass spectrometry of large biomolecules. *Science* 1989;246:64-71.
23. Westermeier R. *Electrophoresis in Practice*. Third edition. Wiley-VCH, Weinheim 2001;81-100.
24. Gygi SP, Corthals GL, Zhang Y, Rochon Y, Aebersold R. Evaluation of two-dimensional gel electrophoresis-based proteome analysis technology. *Proc Natl Acad Sci USA* 2000;97:9390-5.
25. Vuong GL, Weiss SM, Kammer W, Priemer M, Vingron M, Nordheim A, Cahill MA. Improved sensitivity proteomics by postharvest alkylation and radioactive labelling of proteins. *Electrophoresis* 2000;21:2594-605.
26. Hoving S, Voshol H, van Oostrum J. Towards high performance two-dimensional gel electrophoresis using ultrazoom gels. *Electrophoresis* 2000;21:2617-21.
27. Corthals GL, Wasinger VC, Hochstrasser DF, Sanchez JC. The dynamic range of protein expression: a challenge for proteomic research. *Electrophoresis* 2000;21:1104-15.
28. Yates 3rd JR, Gilchrist A, Howell KE, Bergeron JJ. Proteomics of organelles and large cellular structures. *Nature* 2005;6:702-14.
29. Rabilloud T. Two-dimensional gel electrophoresis in proteomics: old, old fashioned, but it still climbs up the mountains. *Proteomics* 2002;2:3-10.
30. Eng JK, McCormack AL, Yates 3rd JR. An approach to correlate tandem mass spectral data of peptides with amino acid sequences in a protein database. *J Am Soc Mass Spectrom* 1994;5:976-89.
31. McCormack AL, Schieltz DM, Goode B, Yang S, Barnes G, Drubin D, Yates 3rd JR. Direct analysis and identification of proteins in mixtures by LC/MS/MS and database searching at the low-femtomol level. *Anal Chem* 1997;69:767-76.
32. Liu H, Sadygov RG, Yates 3rd JR. A model for random sampling and estimation of relative protein abundance in shotgun proteomics. *Anal Chem* 2004;76:4193-201.
33. Link AJ, Eng J, Schieltz DM, Carmack E, Mize GJ, Morris DR, Garvick BM, Yates 3rd JR. Direct analysis of protein complexes using mass spectrometry. *Nat Biotechnol* 1999;17:676-82.
34. Washburn MP, Wolters D, Yates 3rd JR. Large-scale analysis of the yeast proteome by multidimensional protein identification technology. *Nat Biotechnol* 2001;19:242-7.
35. Ünlü M, Morgan ME, Minden JS. Difference gel electrophoresis: a single gel method for detecting changes in protein extracts. *Electrophoresis* 1997;18:2071-7.
36. Gygi SP, Rist B, Gerber SA, Turecek F, Gelb MH, Aebersold R. Quantitative analysis of complex protein mixtures using isotope-coded affinity tags. *Nat Biotechnol* 1999;17:994-9.
37. Davie EW, Neurath H. Identification of a peptide released during autocatalytic activation of trypsinogen. *J Biol Chem* 1955;212:515-29.
38. Macfarlane RG. An enzyme cascade in the blood clotting mechanism, and its function as a biochemical amplifier. *Nature* 1964;202:498-9.

39. Davie EW, Ratnoff OD. Waterfall sequence for intrinsic blood clotting. *Science* 1964;145:1310-2.
40. Black RA, Rauch CT, Kozlosky CJ, Peschon JJ, Slack JL, Wolfson MF, Castner BJ, Stocking KL, Reddy P, Srinivasan S, Nelson N, Boiani N, Schooley KA, Gerhart M, Davis R, Fitzner JN, Johnson RS, Paxton RJ, March CJ, Cerretti DP. A metalloproteinase disintegrin that releases tumour-necrosis factor-alpha from cells. *Nature* 1997;385:729-33.
41. Moss ML, Jin SL, Milla ME, Bickett DM, Burkhart W, Carter HL, Chen WJ, Clay WC, Didsbury JR, Hassler D, Hoffman CR, Kost TA, Lambert MH, Leesnitzer MA, McCauley P, McGeehan G, Mitchell J, Moyer M, Pahel G, Rocque W, Overton LK, Schoenen F, Seaton T, Su JL, Becherer JD. Cloning of a disintegrin metalloproteinase that processes precursor tumour-necrosis factor-alpha. *Nature* 1997;385:733-6.
42. Peschon JJ, Slack JL, Reddy P, Stocking KL, Sunnarborg SW, Lee DC, Russell WE, Castner BJ, Johnson RS, Fitzner JN, Boyce RW, Nelson N, Kozlosky CJ, Wolfson MF, Rauch CT, Cerretti DP, Paxton RJ, March CJ, Black RA. An essential role for ectodomain shedding in mammalian development. *Science* 1998;282:1281-4.
43. McQuibban GA, Gong JH, Tam EM, McCulloch CA, Clark-Lewis I, Overall CM. Inflammation dampened by gelatinase A cleavage of monocyte chemoattractant protein-3. *Science* 2000;289:1202-6.
44. Parks WC, Wilson CL, Lopez-Boada YS. Matrix metalloproteinases as modulators of inflammation and innate immunity. *Nat Rev Immunol* 2004;4:617-29.
45. Hintermann E, Quaranta V. Epithelial cell motility on laminin-5: regulation by matrix assembly, proteolysis, integrins and erbB receptors. *Matrix Biol* 2004;23:75-85.
46. O'Reilly MS, Holmgren L, Chen C, Folkman J. Angiostatin induces and sustains dormancy of human primary tumors in mice. *Nature Med* 1996;2:689-92.
47. Bergers G, Javaherian K, Lo KM, Folkman J, Hanahan D. Effects of angiogenesis inhibitors on multistage carcinogenesis in mice. *Science* 1999;284:808-12.
48. Overall CM, Blobel CP. In search of partners: linking extracellular proteases to substrates. *Nat Rev Mol Cell Biol* 2007;8:245-257.
49. Overall CM, Kleifeld O. Validating MMPs as drug targets and anti-targets for cancer therapy. *Nature Rev Cancer* 2006;6:227-39.
50. Lopez-Otin C, Overall CM. Protease degradomics: a new challenge for proteomics. *Nat Rev Mol Cell Biol* 2002;3:509-19.
51. Tam E.M, Morrison CJ, Wu YI, Stack MS, Overall CM. Membrane protease proteomics: Isotope-coded affinity tag MS identification of undescribed MT1-matrix metalloproteinase substrates. *Proc Natl Acad Sci U S A* 2004;101:6917-22.
52. Hwang IK, Park SM, Kim SY, Lee ST. A proteomic approach to identify substrates of matrix metalloproteinase-14 in human plasma. *Biochim Biophys Acta* 2004;1702:79-87.
53. Anderson NL, Anderson NG. The human plasma proteome: history, character, and diagnostic prospects. *Mol Cell Proteomics* 2002;1:845-67.
54. Guo L, Eisenman JR, Mahimkar RM, Peschon JJ, Paxton RJ, Black RA, Johnson RS. A proteomic approach for the identification of cell-surface proteins shed by metalloproteases. *Mol Cell Proteomics* 2002;130-6.

55. Bech-Serra JJ, Santiago-Josefat B, Esselens C, Saftig P, Baselga J, Arribas J, Canals F. Proteomic identification of desmoglein 2 and activated leukocyte cell adhesion molecule as substrates of ADAM17 and ADAM10 by difference gel electrophoresis. *Mol Cell Biol* 2006;26:5086-95.
56. Overall CM, Dean RA. Degradomics: systems biology of the protease web. Pleiotropic roles of MMPs in cancer. *Cancer Metastasis Rev* 2006;25:69-75.
57. Overall CM. Dilating the degradome: Matrix metalloproteinase 2 (MMP-2) cuts to the heart of the matter. *Biochem J* 2004;383:e5-e7.
58. Parks WC, Wilson CL, Lopez-Boado YS. Matrix metalloproteinases as modulators of inflammation and innate immunity. *Nat Rev Immunol* 2004;4:617-29.
59. Titani K, Torff HJ, Hormel S, Kumar S, Walsh KA, Rodl J, Neurath H, Zwilling R. Amino acid sequence of a unique protease from the crayfish Astacus fluviatilis. *Biochemistry* 1987;26:222-6.
60. Stöcker W, Sauer B, Zwilling R. Kinetics of nitroanilide cleavage by astacin. *Biol Chem Hoppe Seyler* 1991;372:385-92.
61. Dumermuth E, Sterchi EE, Jiang WP, Wolz RL, Bond JS, Flannery AV, Beynon RJ. The astacin family of metalloendopeptidases. *J Biol Chem* 1991;266:21381-5.
62. Stöcker W, Zwilling R. Astacin. *Methods Enzymol* 1995;248:305-25.
63. Gomis-Rüth FX, Stöcker W, Huber R, Zwilling R, Bode W. Refined 1.8 A X-ray crystal structure of astacin, a zinc-endopeptidase from the crayfish Astacus astacus L. Structure determination, refinement, molecular structure and comparison with thermolysin. *J Mol Biol* 1993;229:945-68.
64. Stöcker W, Gomis-Rüth FX, Bode W, Zwilling R. Implications of the three-dimensional structure of astacin for the structure and function of the astacin family of zinc-endopeptidases. *Eur J Biochem* 1993;214:215-31.
65. Grams F, Dive V, Yiotakis A, Yiallouros I, Vassiliou S, Zwilling R, Bode W, Stöcker W. Structure of astacin with a transition-state analogue inhibitor. *Nat Struct Biol* 1996;3:671-5.
66. Bond JS, Beynon RJ. The astacin family of metalloendopeptidases. *Protein Sci* 1995;4:1247-61.
67. Yasumasu S, Katow S, Hamazaki TS, Iuchi I, Yamagami K. Two constituent proteases of a teleostean hatching enzyme: concurrent syntheses and packaging in the same secretory granules in discrete arrangement. *Dev Biol* 1992;149:349-56.
68. Lee KS, Yasumasu S, Nomura K, Iuchi I. HCE, a constituent of the hatching enzymes of Oryzias latipes embryos, releases unique proline-rich polypeptides from its natural substrate, the hardened chorion. *FEBS Lett* 1994;339:281-4.
69. Yan L, Pollock GH, Nagase H, Sarras MP Jr. A 25.7 x 10 (3) M (r) hydra metalloproteinase (HMP1), a member of the astacin family, localizes to the extracellular matrix of Hydra vulgaris in a head-specific manner and has a developmental function. *Development* 1995;121:1591-602.
70. Yan L, Leontovich A, Fei K, Sarras MP Jr. Hydra metalloproteinase 1: a secreted astacin metalloproteinase whose apical axis expression is differentially regulated during head regeneration. *Dev Biol* 2000;219:115-28.
71. Wozney JM, Rosen V, Celeste AJ, Mitsock LM, Whitters MJ, Kriz RW, Hewick RM, Wang EA. Novel regulators of bone formation: molecular clones and activities. *Science* 1988;242:1528-34.
72. Kessler E, Takahara K, Biniaminov L, Brusel M, Greenspan DS. Bone morphogenetic protein-1: the type I procollagen C-proteinase. *Science* 1996;271:360-2.

73. Ferguson EL, Anderson KV. Decapentaplegic acts as a morphogen to organize dorsal-ventral pattern in the Drosophila embryo. *Cell* 1992;71:451-61.
74. Jiang W, Gorbea CM, Flannery AV, Beynon RJ, Grant GA, Bond JS. The alpha subunit of meprin A. Molecular cloning and sequencing, differential expression in inbred mouse strains, and evidence for divergent evolution of the alpha and beta subunits. *J Biol Chem* 1992;267:9185-93.
75. Corbeil D, Gaudoux F, Wainwright S, Ingram J, Kenny AJ, Bioleau G, Crine P. Molecular cloning of the alpha-subunit of rat endopeptidase-24.18 (endopeptidase-2) and co-localization with endopeptidase-24.11 in rat kidney by in situ hybridization. *FEBS Lett* 1992;309:203-8.
76. Johnson GD, Hersh LB. Cloning a rat meprin cDNA reveals the enzyme is a heterodimer. *J Biol Chem* 1992;267:13505-12.
77. Gorbea CM, Marchand P, Jiang W, Copeland NG, Gilbert DJ, Jenkins NA, Bond JS. Cloning, expression, and chromosomal localization of the mouse meprin beta subunit. *J Biol Chem* 1993;268:21035-43.
78. Dumermuth E, Eldering JA, Grünberg J, Jiang W, Sterchi EE. Cloning of the PABA peptide hydrolase alpha subunit (PPH alpha) form human small intestine and its expression in COS-1 cells. *FEBS Lett* 1993;335:367-75.
79. Eldering JA, Grünberg J, Hahn D, Croes HJ, Fransen JA, Sterchi EE. Polarised expression of human intestinal N-benzoyl-L-tyrosyl-p-aminobenzoic acid hydrolase (human meprin) alpha and beta subunits in Madin-Darby canine kidney cells. *Eur J Biochem* 1997;247:920-32.
80. Sterchi EE, Green JR, Lentze MJ. Non-pancreatic hydrolysis of N-benzoyl-l-tyrosyl-p-aminobenzoic acid (PABA-peptide) in the human small intestine. *Clin Sci (Lond)* 1982;62:557-60.
81. Imondi AR, Stradley RP, Wolgemuth R. Synthetic peptides in the diagnosis of exocrine pancreatic insufficiency in animals. *Gut* 1972;13:726-31.
82. Gyr K, Stalder GA, Schiffmann I, Fehr C, Vonderschmitt D, Fahrlaender H. Oral administration of a chymotrypsin-labile peptide - a new test of exocrine pancreatic function in man (PFT). *Gut* 1976;17:27-32.
83. Bornschein W, Goldmann FL, Otte M. [Methodological and first clinical investigation using a new pancreatic function test]. *Clin Chim Acta* 1976;67:21-7.
84. Beynon RJ, Shannon JD, Bond JS. Purification and characterization of a metallo-endoproteinase from mouse kidney. *Biochem J* 1981;199:591-8.
85. Kenny AJ, Fulcher IS, Ridgwell K, Ingram J. Microvillar membrane neutral endopeptidases. *Acta Biol Med Ger* 1981;40:1465-71.
86. Kenny AJ, Ingram J. Proteins of the kidney microvillar membrane. Purification and properties of the phosphoramidon-insensitive endopeptidase ('endopeptidase-2') from rat kidney. *Biochem J* 1987;245:515-24.
87. Grünberg J, Dumermuth E, Eldering JA, Sterchi EE. Expression of the alpha subunit of PABA peptide hydrolase (EC 2.4.24.18) in MDCK cells. Synthesis and secretion of an enzymatically inactive homodimer. *FEBS Lett* 1993;335:376-9.
88. Sterchi EE, Naim HY, Lentze MJ, Hauri HP, Fransen JA. N-benzoyl-L-tyrosyl-p-aminobenzoic acid hydrolase: a metalloendopeptidase of the human intestinal microvillus membrane which degrades biologically active peptides. *Arch Biochem Biophys* 1988;265: 105-118.

89. Hahn D, Lottaz D, Sterchi EE. C-cytosolic and transmembrane domains of the N-benzoyl-L-tyrosyl-p-aminobenzoic acid hydrolase alpha subunit (human meprin alpha) are essential for its retention in the endoplasmic reticulum and C-terminal processing. *Eur J Biochem* 1997;247:933-41.
90. Hahn D, Pischitzis A, Rösmann S, Hansen MK, Leuenberger B, Luginbühl U, Sterchi EE. Phorbol 12-myristate 13-acetate-induced ectodomain shedding and phosphorylation of the human meprinbeta metalloprotease. *J Biol Chem* 2003;278:42829-39.
91. Corbeil D, Milhiet PE, Simon V, Ingram J, Kenny AJ, Boilbeau G, Crine P. Rat endopeptidase-24.18 alpha subunit is secreted into the culture medium as a zymogen when expressed by COS-1 cells. *FEBS Lett* 1993;335:361-6.
92. Johnson GD, Hersh LB. Expression of meprin subunit precursors. Membrane anchoring through the beta subunit and mechanism of zymogen activation. *J Biol Chem* 1994;269:7682-8.
93. Sterchi EE, Naim HY, Lentze MJ. Biosynthesis of N-benzoyl-L-tyrosyl-p-aminobenzoic acid hydrolase: disulfide-linked dimers are formed at the site of synthesis in the rough endoplasmic reticulum. *Arch Biochem Biophys* 1988;265:199-27.
94. Marchand P, Tang J, Johnson GD, Bond JS. COOH-terminal proteolytic processing of secreted and membrane forms of the alpha subunit of the metalloprotease meprin A. Requirement of the I domain for processing in the endoplasmic reticulum. *J Biol Chem* 1995;270:5449-56.
95. Lottaz D, Hahn D, Müller S, Müller C, Sterchi EE. Secretion of human meprin from intestinal epithelial cells depends on differential expression of the alpha and the beta subunits. *Eur J Biochem* 1999;259:496-504.
96. Matters GL, Bond JS. Expression and regulation of the meprin beta gene in human cancer cells. *Mol Carcinog* 1999;25:169-78.
97. Trachtman H, Valderrama E, Dietrich JM, Bond JS. The role of meprin A in the pathogenesis of acute renal failure. *Biochem Biophys Res Commun* 1995;208:498-505.
98. Lottaz D, Maurer CA, Hahn D, Büchler MW, Sterchi EE. Nonpolarized secretion of human meprin alpha in colorectal cancer generates an increased proteolytic potential in the stroma. *Cancer Res* 1999;59:1127-33.
99. Yamaguchi T, Fukase M, Kido H, Sugimoto T, Katunuma N, Chihara K. Meprin is predominantly involved in parathyroid hormone degradation by the microvillar membranes of rat kidney. *Life Sci* 1994;54:381-6.
100. Bertenshaw GP, Turk BE, Hubbard SJ, Matters GL, Bylander JE, Crisman JM, Cantley LC, Bond JS. Marked differences between metalloproteases meprin A and B in substrate and peptide bond specificity. *J Biol Chem* 2001;276:13248-55.
101. Kaushal GP, Walker PD, Shah SV. An old enzyme with a new function: purification and characterization of a distinct matrix-degrading metalloproteinase in rat kidney cortex and its identification as meprin. *J Cell Biol* 1994;126:1319-27.
102. Walker PD, Kaushal GP, Shah SV. Meprin A, the major matrix degrading enzyme in renal tubules, produces a novel nidogen fragment in vitro and in vivo. *Kidney Int* 1998;53:1673-80.
103. Villa JP, Bertenshaw GP, Bylander JE, Bond JS. Meprin proteolytic complexes at the cell surface and in extracellular spaces. *Biochem Soc Symp* 2003;53-63.

104. Herzog, C., Kaushal, G. P., Haun, R. S. Generation of biologically active interleukin-1beta by meprin B. *Cytokine* 2005;31:394-403.

# Chapter 2

## 2. Sample preparation of culture medium from Madin-Darby canine kidney cells[1]

### 2.1 Summary

A reproducible, standardized and simple sample preparation methodology is the key to successful two-dimensional gel electrophoresis. This chapter describes step-by-step the sample preparation of culture medium from Madin-Darby canine kidney cells. Tips and tricks are given to circumvent possible pitfalls.

---

[1] This book chapter has been published previously in: "Ambort D, Lottaz D and Sterchi EE (2008) Sample preparation of culture medium from Madin-Darby canine kidney cells. In *2D PAGE: Sample Preparation and Fractionation Volume 2* (Posch A, ed), chapter 10, pp. 113-130. Humana Press, Totowa, NJ."

## 2.2 Introduction

2-DE, introduced by O'Farrell and Klose in 1975 (1, 2), enabled separation of complex protein mixtures into individual protein species according to their net charge (p$I$) in the first dimension by IEF and in the second dimension according to their molecular mass ($M_r$) by SDS-PAGE (3). In the conventional approach, IEF was performed in carrier ampholyte-generated pH gradients, which moved towards the cathode upon prolonged focusing time. This „cathodic drift" phenomenon was thereafter remedied by non-equilibrium pH gradient gel electrophoresis (4) and finally eliminated with the invention of IPGs (5-8). The development of microanalytical techniques, namely Edman sequencing (9-11) and mass spectrometry (12-14) enabled identification of proteins at amounts available from a single 2-D gel. 2-DE advanced to the core technology of proteome analysis (7, 8, 15) and was brought from art to craft in an industrial standard.

Appropriate sample treatment is the key to good results. The ideal sample solubilization procedure should result in the disruption of all non-covalently bound protein complexes and aggregates into a solution of individual polypeptides (15). Denaturation and reduction of proteins is achieved in the standard lysis buffer (O'Farrell 1975) (1) which is composed of 8-9 M urea, 2-4% CHAPS, 1% DTT or DTE and 0.8-2% carrier ampholytes. Hydrophobic proteins are better dissolved in 2 M thiourea and 7 M urea instead of 9 M urea (16). Optimized procedures for different sample types do exist (17). However, a general „Prepares them all" procedure is not available (18). Another high-priority issue is the removal and inactivation of all interfering substances: nucleic acids, lipids, salts, small ionic compounds, polysaccharides, proteases and insoluble particles. Sample preparation for 2-DE is a very cumbersome and time-consuming task that is subject to trial and error.

Due to the complex biological architecture of eukaryotic cells into organelles and large cellular structures, fractionation techniques are applied before comprehensively studying the subproteomes. In such reductionist approaches classic biochemical techniques, namely centrifugation and affinity-mediated isolation using antibodies against molecular tags, are applied in order to enrich for subcellular fractions as reviewed by Yates 3rd (19). Beside analysis of cytosolic, organelle-specific and transmembrane proteins several investigations were aimed at identifying those proteins secreted by various cell types into the extracellular milieu or medium (20-24).

This chapter describes the sample preparation of culture medium from MDCK cells for 2-DE (Fig. 1). The methodology is sub-sectioned into four parts with basic introductory information on each topic: cell culture (*see* Subheading 2.4.1), ultracentrifugation and ultrafiltration (*see* Subheading 2.4.2), protein quantitation by BCA assay (*see* Subheading 2.4.3) and finally, rehydration loading and isoelectric focusing (*see* Subheading 2.4.4).

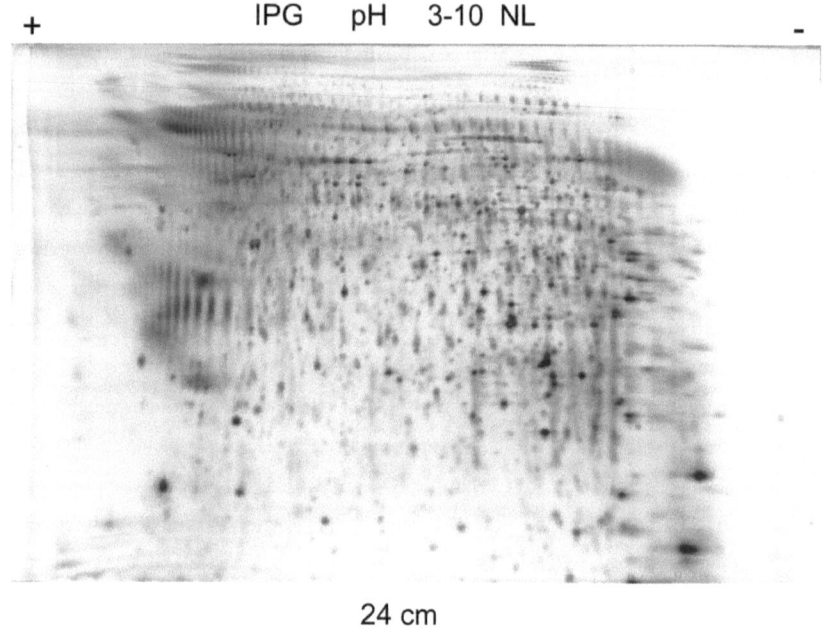

**Fig. 1. Two-dimensional gel electrophoresis of Madine-Darby canine kidney cell culture supernatant (80 µg protein).**

First dimension: isoelectric focusing in an immobilized pH gradient (IPG) pH 3-10 non-linear (NL) in a 24 cm long gel strip. Second dimension: SDS-PAGE in a 12.5% gel. Silver stained.

## 2.3 Materials

## 2.3.1 Cell culture

### 2.3.1.1 Equipment

1. BD Falcon™ bulk packaged serological pipets (10 mL) (BD Biosciences, Franklin Lakes, NY, USA)
2. BD Falcon™ individually wrapped serological pipets (25 mL) (BD Biosciences, Franklin Lakes, NY, USA)
3. BD Falcon™ standard cell culture dish, standard tissue-culture treated (100 x 20 mm) (BD Biosciences, Franklin Lakes, NY, USA)
4. CELLSTAR® PP-test tubes (50 mL, sterile) (Greiner Bio-One Inc., Longwood, FL, USA)
5. Erlenmeyer flasks (250 mL)
6. Laminar air flow cabinet (Brouwer AG, Luzern, Switzerland)
7. Neubauer improved counting chamber (Assistent, Sondheim, Germany)
8. NUAIRE™ US autoflow $CO_2$ water-jacketed incubator NU-4750 (Vitaris AG, Baar, Switzerland)
9. Water bath

### 2.3.1.2. Solutions and reagents

1. Dulbecco's Phosphate Buffered Saline (D-PBS) (1X) (500 mL) (GIBCO Invitrogen corporation, Grand Island, NY, USA)
2. Foetal Bovine Serum (FBS) (E. U. approved South American origin, virus and mycoplasma tested) (500 mL) (GIBCO Invitrogen corporation, Grand Island, NY, USA)
3. Minimum Essential Medium (MEM) (1X) (with Earle's salts, without L-glutamine) (500 mL) (GIBCO Invitrogen corporation, Grand Island, NY, USA)
4. Penicillin-Streptomycin-Glutamine (100X) (100 mL) (GIBCO Invitrogen corporation, Grand Island, NY, USA)
5. Trypsin-EDTA (1X) (100 mL) (GIBCO Invitrogen corporation, Grand Island, NY, USA)
6. Culture medium: 1X MEM (*see* Note 1), 5% (v/v) FBS, 100 units/mL penicillin, 100 µg/mL streptomycin and 292 µg/mL L-glutamine. To 500 mL of MEM (one bottle) aseptically add 25 mL of FBS (*see* Note 2) and 5 mL of 100X Penicillin-Streptomycin-Glutamine stock solution. Store at 4°C. Before use warm up to 37°C in a water bath.
7. Serum-free medium: 1X MEM (*see* Note 1), 100 units/mL penicillin, 100 µg/mL streptomycin and 292 µg/mL L-glutamine. To 500 mL of MEM (one bottle) aseptically add 5 mL of 100X Penicillin-Streptomycin-Glutamine stock solution. Store at 4°C. Before use warm up to 37°C in a water bath.

## 2.3.2 Ultracentrifugation and ultrafiltration

### 2.3.2.1 Equipment

1. Centricon® Plus-70 centrifugal filter devices (Millipore corporation, Billerica, MA, USA)
2. Centrifuge filter system (50 mL, 0.2 μm) (Costarcorporation, Cambridge, MA, USA)
3. Eppendorf centrifuge 5415R (Eppendorf AG, Hamburg, Germany)
4. Eppendorf tubes (1.5 mL, 2 mL)
5. KONTRON CENTRIKON TFT 70.38 fixed-angle rotor (KONTRON Instruments AG, Zürich, Switzerland)
6. KONTRON CENTRIKON T-2060 ultracentrifuge (KONTRON Instruments AG, Zürich, Switzerland)
7. KONTRON CENTRIKON ultracentrifuge tubes (32.5 mL) (KONTRON Instruments AG, Zürich, Switzerland)
8. Mettler AC 100 analytical balance (Mettler Instrumente AG, Greifensee, Zürich, Switzerland)
9. Sorvall RT6000D centrifuge (Kendro Laboratory Products AG, Zürich, Switzerland)
10. Sorvall H1000B swinging bucket rotor (Kendro Laboratory Products AG, Zürich, Switzerland)
11. Water bath

### 2.3.2.2 Solutions and reagents

1. Ethylenedinitrilo tetraacetic acid disodium salt dihydrate ($Na_2$-EDTA $2H_2O$, Titriplex® III) (GR for analysis) (Merck, Darmstadt, Germany)
2. Phenylmethylsulfonyl fluoride (PMSF) (Sigma, St. Louis, MO, USA)
3. 2-Propanol (GR for analysis) (Merck, Darmstadt, Germany)
4. Sodium hydroxide (NaOH) (pellets GR for analysis) (Merck, Darmstadt, Germany)
5. Tris(hydroxymethyl)aminomethane (Tris) (GR for analysis buffer substance) (Merck, Darmstadt, Germany)
6. 0.5 M EDTA pH 8.0 stock solution: To make 100 mL of stock solution, dissolve 2 g of NaOH pellets in 80 mL of dd$H_2O$. Add 18.6 g of $Na_2$-EDTA $2H_2O$ under constant stirring at RT (*see* Note 3). Titrate solution to pH 8 with 5 N NaOH (liquid). Adjust to a final volume of 100 mL with dd$H_2O$ and check pH again. Filter solution with a 0.2 μm bottle top filter. This solution can be stored at RT.
7. 0.1 M PMSF stock solution: To prepare 25 mL, dissolve 0.44 g of PMSF in 25 mL of 2-propanol (isopropanol). Warm up to 37°C in a water bath (*see* Note 4). Portion solution into 2 mL aliquots in 2 mL Eppendorf tubes and store at –20°C.
8. 0.2 M Tris pH 10.5 stock solution: To make 0.5 L, dissolve 12.12 g of Tris in 0.5 L of dd$H_2O$ (*see* Note 5). Filter solution with a 0.2 μm bottle top filter. This solution can be stored at RT.
9. Sample solubilization buffer I: 20 mM Tris pH 9.0, 1 mM EDTA, 1 mM PMSF. To prepare 150 mL, dilute 15 mL of 0.2 M Tris pH 10.5 stock solution to a final volume of 150 mL in dd$H_2O$. Add 0.3 mL of 0.5 M EDTA pH 8.0 stock solution and 1.5 mL of 0.1 M PMSF stock solution under constant stirring at

RT (*see* Note 6). Filter solution with a 0.2 µm bottle top filter. This solution should be prepared freshly just prior to use!

## 2.3.3 Protein quantitation by BCA assay

### 2.3.3.1 Equipment

1. Beaker (25 mL)
2. CELLSTAR® micro-plate (TC, sterile) (Greiner Bio-One Inc., Longwood, FL, USA)
3. Incubator
4. Vortex mixer
5. Vmax® microplate reader (Molecular Devices Corporation, Sunnyvale, CA, USA)

### 2.3.3.2 Solutions and reagents

1. Albumin Standard Ampules (2 mg/mL, 10 x 1 mL) (Pierce, Rockford, IL, USA)
2. BCA™ Protein Assay Kit (Pierce, Rockford, IL, USA)
3. BCA™ Protein Assay Reagent A (500 mL) (Pierce, Rockford, IL, USA)
4. BCA™ Protein Assay Reagent B (25 mL) (Pierce, Rockford, IL, USA)

## 2.3.4 Rehydration loading and isoelectric focusing

### 2.3.4.1 Equipment

1. Beaker (50 mL)
2. Centrifuge filter system (50 mL, 0.2 µm) (Costarcorporation, Cambridge, MA, USA)
3. Eppendorf centrifuge 5415R (Eppendorf AG, Hamburg, Germany)
4. EPS 3501 XL power supply (Amersham Biosciences, Uppsala, Sweden)
5. IEF electrode strips (Amersham Biosciences, Uppsala, Sweden)
6. Immobiline™ DryStrip Kit (Amersham Biosciences, Uppsala, Sweden)
7. Immobiline™ DryStrip Reswelling Tray (7-24 cm) (Amersham Biosciences, Uppsala, Sweden)
8. Vortex mixer
9. Multiphor™ II horizontal electrophoresis apparatus (Amersham Biosciences, Uppsala, Sweden)
10. Multitemp™ III thermostatic circulator (Amersham Biosciences, Uppsala, Sweden)
11. Multi-purpose rotator (Scientific Industries Inc., Queens Village, NY, USA)
12. Parafilm (50 cm x 15 m) (American National Can Company, Chicago, IL, USA)
13. Petri dishes
14. Tweezers

*2.3.4.2 Solutions and reagents*

1. Bromophenol blue (BPB) (Bio-Rad Laboratories, Richmond, CA, USA)
2. 3-[(3-cholamidopropyl)-dimethylammonio]-1-propane sulfonate (CHAPS) (ultrapure) (USB corporation, Cleveland, OH, USA)
3. 1,4-dithioerythritol (DTE) (for biochemistry) (Merck, Darmstadt, Germany)
4. Immobiline™ DryStrip pH 3-10 NL (IPG) (240 mm x 3 mm x 0.5 mm) (Amersham Biosciences, Uppsala, Sweden)
5. Mixed bed ion exchanger resin
6. Paraffin (Merck, Darmstadt, Germany)
7. Pharmalyte™ 3-10 (for IEF) (Amersham Biosciences, Uppsala, Sweden)
8. Thiourea (puriss. p. a. ACS; ≥ 99% (RT)) (Fluka, Buchs, Switzerland)
9. ZOOM® urea (Invitrogen life technologies, Carlsbad, CA, USA)
10. DTE aliquots: 65 mM in 1.5 mL of sample solubilization buffer II (working solution). Portion 0.015 g (15 mg) of DTE in 1.5 mL Eppendorf tube and store at 4°C until use (*see* Note 7).
11. Sample solubilization buffer II (stock solution): 7 M urea, 2 M thiourea, 4% (w/v) CHAPS. To prepare 25 mL, dissolve 10.5 g of urea, 3.8 g of thiourea in 10 mL of ddH$_2$O under constant stirring at RT. Fill up to a final volume of 30 mL with ddH$_2$O (*see* Note 8). Add 5 g of mixed bed ion exchanger resin and stir for 10 minutes. Remove beads by filtration through a 0.2 µm bottle top filter. Dissolve 1 g of CHAPS, add a trace of BPB and portion solution into 1.5 mL aliquots in 1.5 mL Eppendorf tubes. Store at –20°C until use.
12. Sample solubilization buffer II (working solution): 7 M urea, 2 M thiourea, 4% (w/v) CHAPS, 1% (w/v) (65 mM) DTE, 2% (v/v) (0.8% (w/v)) Pharmalyte 3-10. To make up 1.5 mL, add 1.5 mL of sample solubilization buffer II (stock solution) to one DTE aliquot in 1.5 mL Eppendorf tube. Add 30 µL of Pharmalyte 3-10 and incubate for 15 minutes at RT with occasional vortexing until DTE is completely dissolved. This solution is prepared just prior to use (*see* Note 9).

## 2.4 Methods

### 2.4.1 Cell culture

Mammalian renal tubular epithelium consists of at least seven different segments, complicating biochemical investigation of this heterogeneous tissue. Cultured monolayers of dog kidney (MDCK) cells display many typical features of renal tubular epithelia, such as brush border membrane, tight junctions and adherent junctions. MDCK strain II (26) was used to prepare samples of cell culture supernatants for 2-DE. High-quality protein samples for 2-DE are only obtained from serum-free media. Serum-derived proteins heavily contaminate cell-derived secreted proteins in culture media (*see* Note 10). Hence it is essential to wash the cells thoroughly (twice in PBS or serum-free medium) when changing from complete culture media to serum-free conditions.

1. Harvest confluent MDCK cells by trypsinization. To five confluent 100-mm cell culture dishes add 1.5 mL of prewarmed (37°C) Trypsin-EDTA solution per dish and incubate at 37°C in a humidified incubator in an atmosphere of 5% $CO_2$ until cells detach (*see* Note 11).
2. Resuspend trypsinized cells in culture medium. To each dish add 6.5-7 mL of prewarmed (37°C) culture medium and pool resuspended cells from the five dishes into one 50 mL Greiner tube. Fill up to a total volume of 40 mL with culture medium (*see* Note 12).
3. Seed $1.15 \times 10^6$ cells onto 100-mm cell culture dishes. Adjust final volume with prewarmed (37°C) culture medium to 9 mL per dish. Prepare a total of 18 dishes per condition (*see* Note 13). Incubate the cultures for about three days at 37°C in a humidified incubator in an atmosphere of 5% $CO_2$ until cells are confluent (*see* Note 14).
4. Aspirate medium and wash cells twice in 4 mL of prewarmed (37°C) serum-free medium per dish (*see* Note 15). Add 4 mL of serum-free medium to each dish and incubate for 22 hours at 37°C in a humidified incubator in an atmosphere of 5% $CO_2$.
5. Harvest cell culture supernatants. Pool media from 18 dishes into one 250 mL Erlenmeyer flask to give a final volume of 70-72 mL (*see* Note 13). Immediately proceed to ultracentrifugation and ultrafiltration (*see* Subheading 2.4.2).

## 2.4.2 Ultracentrifugation and ultrafiltration

Beside body fluids, such as human plasma, urine and cerebrospinal fluid, cell culture supernatants are among the most difficult samples to be prepared for 2-D-PAGE. Culture media contain the complete set of interfering substances that are incompatible with the first dimensional isoelectric focusing: insoluble particles (dead cells, cell debris), nucleic acids (from dead cells), lipids (membranes and exosomes (22)), salts (from medium *see* Note 1), small ionic compounds (amino acids from medium), phenolic compounds (Phenol red from medium) and proteases (secreted during cultivation). Therefore these contaminants are removed in a two-step purification strategy: 1) insoluble particles, nucleic acids and lipids by high-speed centrifugation; 2) salts, small ionic and phenolic compounds by ultrafiltration. Proteases are inhibited by addition of inhibitors and basic pH conditions. Alternative methods such as TCA/acetone precipitation (27) and dialysis must not be applied. Unspecific salt precipitation and loss of proteins are associated with these methods (*see* Note 16).

1. Add 140 µL of 0.5 M EDTA pH 8.0 and 700 µL of 0.1 M PMSF to 70 mL of culture medium in 250 mL Erlenmeyer flask and put on ice (*see* Note 17).
2. Transfer 4 x 17.5 mL of medium to four prechilled KONTRON CENTRIKON ultracentrifuge tubes. Adjust precise volumes with ddH$_2$O on an analytical balance and put tubes on ice.
3. Place precooled (4°C) KONTRON CENTRIKON TFT 70.38 fixed-angle rotor into KONTRON CENTRIKON T-2060 ultracentrifuge. Place tubes into the rotor in appropriate positions and close the lid. Ultracentrifuge for 60 min at 31,200 rpm (100,000$g$) at 4°C.
4. Carefully remove tubes from ultracentrifuge and put on ice.
5. Rinse Centricon® Plus-70 components consisting of cap, concentrate/retentate cup, sample filter cup and filtrate collection cup with ddH$_2$O to remove dust particles (*see* Note 18).
6. Place sample filter cup into filtrate collection cup and leave on ice.
7. Pool the medium by inverting tubes into the same sample filter cup and close. Then place assembled Centricon® Plus-70 centrifugal filter device into precooled (4°C) Sorvall H1000B swinging bucket rotor fixed in a Sorvall RT6000D centrifuge. Centrifuge for 60 min at 3,200 rpm (2,190$g$) at 4°C.
8. Discard the flow-through in the filtrate collection cup. Add 70 mL of prechilled sample solubilization buffer I (20 mM Tris pH 9.0, 1 mM EDTA, 1 mM PMSF) into sample filter cup (*see* Note 19). Centrifuge for 60 min at 3,200 rpm (2,190$g$) at 4°C.
9. Repeat step 8 twice.
10. After the last washing step place concentrate/retentate cup upside down onto the filtrate collection cup. Invert assembly and place back into centrifuge. Recover sample concentrate for 5 min at 2,200 rpm (1,000$g$) at 4°C.
11. Determine volume with a pipette. Typical final concentrate volumes are between 250 µL and 350 µL.

12. Transfer protein concentrate to a 1.5 mL Eppendorf tube and spin for 5 min at 13,200 rpm (16,100$g$) at 4°C to remove precipitates (*see* Note 20). Put samples on ice and proceed to protein quantitation (*see* Subheading 2.4.3) or store at –20°C until use.

## *2.4.3 Protein quantitation by BCA assay*

Quantitative determination of protein solubilized in modified lysis buffer (16) (7 M urea, 2 M thiourea, 4% CHAPS, 65 mM DTE and 2% Pharmalyte 3-10) is not possible. The Bradford assay (29) cannot be used for two reasons: 1) Coomassie Brilliant Blue G-250 binds to detergents (CHAPS) and carrier ampholytes (Pharmalyte 3-10); 2) Coomassie Brilliant Blue G-250 may not bind to protein at all under basic pH conditions (urea). The second problem may be remedied by acidification of sample with 0.1 N HCl prior to quantitation (30). Standard Lowry (31), Biuret (32) and BCA (bicinchoninic acid) (33) assays based on the reduction of $Cu^{2+}$ to $Cu^+$ for development of colour interfere with thiol reducing agents (DTE) and thiourea. Thiourea forms complexes with copper ions. TCA/acetone precipitation (27) must not be applied in combination with any of these techniques. Protein may be lost upon precipitation leading to underestimation of solubilized protein. The best solution to all problems mentioned above is quantitation of protein prior to solubilization in lysis buffer. In this section the BCA assay from Pierce is used to accurately quantitate protein concentration in sample solubilization buffer I (20 mM Tris pH 9.0, 1 mM EDTA, 1 mM PMSF). Although the BCA assay interferes with chelating agents (EDTA) concentrations below 10 mM are tolerated (34).

1. Prepare a set of albumin (BSA) standards in 1.5 mL Eppendorf tubes (*see* Table 1 for details). Use dd$H_2O$ as diluent. Gently vortex tubes. There will be sufficient volume for two replications of each diluted standard.
2. Prepare protein samples in 1.5 mL Eppendorf tubes. Dilute 5 µL of each protein concentrate to a final volume of 100 µL in dd$H_2O$ (1:20 dilution). Gently vortex tubes. While performing BCA assay put undiluted protein concentrates on ice (*see* Subheading 2.4.2.12).
3. Pipette 25 µL of each standard or unknown sample replicate into a microplate well. Standards are applied in duplicates, protein samples in triplicates. Use dd$H_2O$ for blanks.
4. Prepare BCA working reagent by mixing 50 parts of BCA$^{TM}$ Reagent A with 1 part of BCA$^{TM}$ Reagent B in a 25 mL beaker. Mix thoroughly (*see* Note 21). For each standard, unknown sample or blank 200 µL of BCA working reagent is required. Include two extra replicates in your calculation.
5. Add 200 µL of BCA working reagent to each well. Gently shake microplate by hand for a few seconds.
6. Incubate microplate for 30 min at 37°C in an incubator.
7. Cool plate to RT and measure absorbance at 550 nm on a microplate reader (*see* Note 22).

8. Subtract the average 550 nm absorbance measurement of the blank replicates from the 550 nm absorbance measurements of all other individual standards and unknown sample replicates.
9. Prepare a standard curve by plotting the average blank-corrected 550 nm absorbance measurement ($A_{550}$, y-axis) for each albumin standard versus its amount (in µg, x-axis) in increasing order (*see* Note 23).
10. Use the standard curve to determine the protein amount of each unknown sample (Fig. 2). The protein concentration in unknown sample is calculated as follows: (protein amount of unknown in µg) / (volume of diluted sample in µL) x (dilution factor)= protein concentration in mg/mL (*see* Note 24). Typical protein concentrations are between 5 mg/mL and 7 mg/mL.
11. Portion undiluted protein concentrates into appropriate aliquots (80 µg for analytical load) in 1.5 mL Eppendorf tubes and store at –20°C until use (*see* Subheading 2.4.4.1).

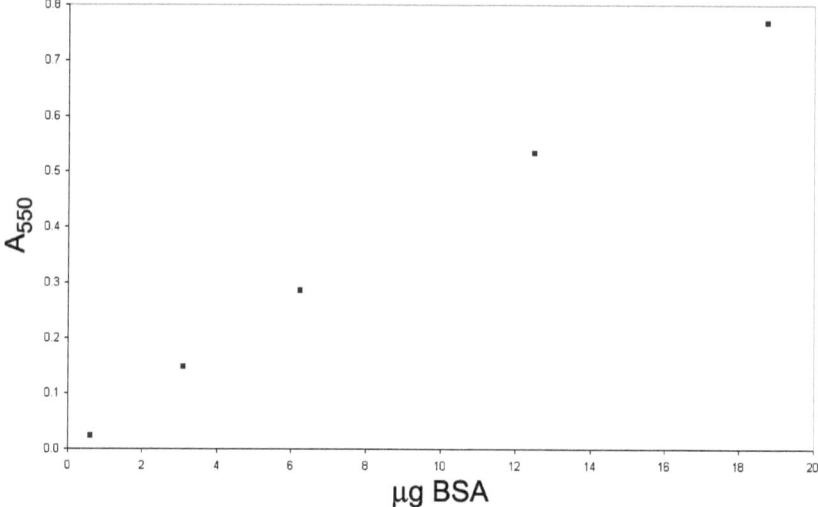

**Fig. 2. Typical color response curve for albumin (BSA) standards using the BCA assay.**
Each point represents the mean of two replications.

**Table 1**

**Preparation of diluted albumin (BSA) standards**

[a] Use ddH$_2$O to prepare albumin (BSA) standards.
[b] The concentration of albumin standard stock solution is 2 mg/mL.

| Vial | Volume of diluent[a] | Volume and source of BSA | Final BSA concentration |
|---|---|---|---|
| A | 150 µL | 150 µL of stock[b] | 1000 µg/mL |
| B | 25 µL | 75 µL of A | 750 µg/mL |
| C | 50 µL | 50 µL of A | 500 µg/mL |
| D | 75 µL | 25 µL of A | 250 µg/mL |
| E | 88 µL | 12.5 µL of A | 125 µg/mL |
| F | 98 µL | 2.5 µL of A | 25 µg/mL |

## *2.4.4 Rehydration loading and isoelectric focusing*

Traditionally, protein samples prepared in standard lysis buffer (O'Farrell 1975) (1) were loaded onto the basic end (cathode) of an IEF tube gel. Before sample loading the gel rods were prerun to establish a carrier ampholyte-derived pH gradient. Upon development of fixed pH gradients samples were applied with rubber frames or sample cups to rehydrated IPG strips at the acidic or basic end (5-8) or simultaneously at both ends (35). The problem with cup-loading sample application techniques is that proteins may precipitate during the sample entry phase, which leads to horizontal streaking at the sample application point. This problem is remedied by in-gel sample rehydration where protein solubilized in lysis buffer is directly diluted with the rehydration solution used for IPG strip reswelling (36, 37). Unfortunately, some proteins that are soluble in lysis buffer may be lost upon dilution into rehydration solution due to lower concentrations of chaotropic agents and detergents. For simplicity protein sample preparation and rehydration can be done all-in-one in modified lysis buffer (16) (7 M urea, 2 M thiourea, 4% CHAPS, 65 mM DTE and 2% Pharmalyte 3-10). This strategy works very well in combination with the Multiphor II horizontal flatbed isoelectric focusing system with final maximum voltage limited to 3500 V for steady-state IEF (38). Higher voltage settings may become problematic, because zwitterionic detergent (CHAPS), reducing agent (DTE) and carrier ampholytes (Pharmalyte 3-10) heavily contribute to the current in the strip.

1. Thaw protein samples on ice (*see* Subheading 2.4.3.11). Dilute each aliquot of protein solution (80 μg for analytical load) to a final volume of 450 μL in sample solubilization buffer II (*see* Note 25).
2. Gently vortex tubes and solubilize protein for 60 min on a rotary shaker at RT.
3. Centrifuge for 30 min at 13,200 rpm (16,100g) at 22°C in a tabletop centrifuge to remove insoluble particles.
4. Slide the protective lid completely off the Immobiline™ DryStrip Reswelling Tray and level the tray by turning the leveling feet until the bubble in the spirit level is centered.
5. Remove IPG strips (240 mm long, 3 mm wide ready-made Immobiline™ DryStrips pH 3-10 NL cast on GelBond PAGfilm) from the freezer and warm up to RT.
6. Evenly apply the entire sample-containing solution into the groove of the reswelling tray (*see* Note 26).
7. Peel off the protective cover sheet from the ready-made IPG strip starting at the acidic (+) end and grip the strip with tweezers at the overlapping basic plastic end (*see* Note 27).
8. Slowly lower the IPG strip (gel side down) onto the solution with the acidic (+) end oriented towards the number labels of the reswelling tray (*see* Note 26).
9. Cover the strip with 3 mL of paraffin oil (*see* Note 28). Repeat steps 6-9 for each sample.
10. Slide the lid onto the reswelling tray and rehydrate the IPG strips overnight at RT.
11. To remove rehydrated IPG strips sequentially from the reswelling tray, open the lid, slide the tip of tweezers along the sloped end of the slot and into the slight depression under the IPG strip. Grab the acidic (+) end of the strip with tweezers and lift the strip out of the tray. While still holding the strip, rinse it briefly with ddH$_2$O. Place it on a piece of damp filter paper at one edge to drain off excess liquid. Repeat procedure for each strip.
12. Set the temperature on the MultiTemp™ III thermostatic circulator to 20° C.
13. Place the ceramic cooling plate in Multiphor II unit and make sure the surface is level.
14. Starting at the top of the plate near the cooling tubes pipette 1 mL of paraffin oil in a straight line onto the middle of the plate. Use the grid of the cooling plate as a guide. Then pipette 1 mL of paraffin oil on each side of the paraffin oil line (a total of 3 mL) onto the bottom of the plate. The additional 1 mL of paraffin oil on each side is evenly spread on the cooling plate to form a triangle which begins with the base line at the bottom and extends to the middle of the paraffin oil line (*see* Note 29).
15. Slowly lower the Immobiline™ DryStrip tray onto the bottom of the paraffin oil triangle with the red (anodic, positively charged) electrode connection of the tray positioned at the top of the plate near the cooling tubes.
16. Connect the red and black electrode leads on the tray to the Multiphor II unit.
17. Pour 10 mL of paraffin oil into the tray at the bottom of the cooling plate.
18. Slowly lower the Immobiline™ DryStrip aligner, 12 grooves side up, onto the bottom of the paraffin oil layer next to the black electrode.
19. Transfer the rehydrated IPG strips with tweezers to adjacent grooves of the aligner in the tray. Place the strips gel side up with the acidic (+) end at the top of the tray near the red electrode.
20. Cut one IEF electrode strip into two pieces each to a length of 110 mm. Moisten the two IEF electrode strips with deionized water (*see* Note 30).
21. Place the damp electrode strips across the acidic and basic ends of the aligned IPG strips.

22. Align each electrode over an electrode strip, ensuring the marked side corresponds to the side of the tray giving electric contact (*see* Note 31). When the electrodes are properly aligned, press them down to contact the electrode strips.
23. Pour 100 mL of paraffin oil into the tray to cover the IPG and electrode strips.
24. Close the lid of the Multiphor II unit. Connect the leads on the lid to the EPS 3501 XL power supply and start IEF according to programmed parameters (*see* Note 32).
25. After IEF remove electrodes and IEF electrode strips.
26. Grip each IPG strip with tweezers at the overlapping basic plastic end, carefully remove it from the tray and rinse it with ddH$_2$O.
27. Place each IPG strip into a petri dish with the plastic side of the strip facing the inner wall of the petri dish and cover. Seal it with a piece of parafilm and store at −20° C until use.
28. IPG strip equilibration, SDS-PAGE and post separation visualization techniques applied following sample preparation and IEF are not topic of this chapter. Useful tips and tricks concerning these methods can be found in the Amersham 2-D electrophoresis handbook (40). As an example the final 2-D map of MDCK cell culture supernatant is shown in Fig. 1.

## 2.5 Notes

1. Composition of Minimum Essential Medium (MEM) in mg/L (25): CaCl$_2$ 2H$_2$O, 264; KCl, 400; MgSO$_4$ 7H$_2$O, 200; NaCl, 6800; NaHCO$_3$ , 2200; NaH$_2$PO$_4$ 2H$_2$O, 158; D-glucose, 1000; Phenol red, 10; L-arginine HCl, 126; L-cystine, 24; L-histidine HCl H$_2$O, 42; L-isoleucine, 52; L-leucine, 52; L-lysine HCl, 73; L-methionine, 15; L-phenylalanine, 32; L-threonine, 48; L-tryptophan, 10; L-tyrosine, 36; L-valine, 46; D-Ca pantothenate, 1; choline chloride, 1; folic acid, 1; i-inositol, 2; niacinamide, 1; pyridoxine HCl, 1; riboflavin, 0.1 and thiamine HCl, 1.
2. Before use Foetal Bovine Serum (FBS) is heat-inactivated! Incubate one bottle (500 mL) of FBS for 30 min at 56°C in a water bath. Portion solution into 25 mL aliquots and store at −20°C.
3. The EDTA may not completely dissolve in 0.5 M stock solution below pH 8.0. Therefore titration of EDTA stock solution with a few drops of 5 N NaOH (liquid) is necessary. The solution is ready when the pale white colour turns into a crystal clear solution.
4. PMSF is very toxic! Protect your eyes and skin. PMSF is not very stable in water and has a half-life of about 30 minutes. Hence the solution is prepared in isopropanol. PMSF is difficult to dissolve at RT; therefore the solution is warmed up to 37°C.
5. The 0.2 M Tris stock solution has a pH of 10.5-10.6. Do not titrate with HCl! The chloride ions extremely contribute to the current (heat production) during IEF. Any heat produced during IEF will cause protein precipitation and produce horizontal streaks in the final 2-D gel.
6. The sample solubilization buffer I has a pH of 9.0-9.1. Do not titrate with HCl! The Tris serves as a positively charged ion that helps in solubilization of proteins and not to maintain constant pH (*see* Note 19). This solution should be prepared shortly before use due to the poor stability of PMSF in water (*see* Note 4). The 0.1 M PMSF stock solution should be warmed up to 37°C in a water bath. Otherwise the PMSF may precipitate!

7. DTE is not very stable in solution. Hence it is stored as solid in small aliquots at 4°C.
8. The final volume of 30 mL compensates for the dead volume of the magnetic stir bar in a 50 mL beaker and equals to a total volume of 25 mL. Add urea and thiourea in small portions with the help of a spatula under constant stirring at RT. Urea and thiourea will cool down the solution and decrease solubility! Therefore in between additions wait for several minutes until each small portion is fully dissolved. Do not heat urea-containing solutions above 37°C to avoid carbamylation of proteins!
9. The sample solubilization buffer II (working solution) should be prepared freshly. Never reuse remaining buffer, better discard it!
10. It is essential to wash the cells thoroughly (twice in PBS or serum-free medium) to remove serum proteins. Culture media heavily contaminated with Foetal Bovine Serum (FBS) resemble human plasma!
11. Before trypsinization confluent cells may be thoroughly washed (twice in 2-3 mL of PBS or serum-free medium) (*see* Note 10). Prewarm PBS and Trypsin-EDTA solution to 37°C in a water bath!
12. To each dish treated with 1.5 mL of Trypsin-EDTA solution 6.5-7 mL of culture medium is added to give a total volume of 40 mL. Prewarm culture medium to 37°C in a water bath!
13. In total 18 dishes per condition are prepared. The final volume of culture medium used per dish is 9 mL. Once MDCK cells reach confluence serum-free medium is added. The final volume of serum-free medium used per dish is 4 mL. This gives a total volume of 70-72 mL per condition and corresponds to the appropriate processing volume for the Centricon® Plus-70 centrifugal filter devices (*see* Note 18).
14. The doubling time of MDCK strain II cells is one day. Confluence is reached after 3-4 days.
15. Alternatively, use PBS instead of serum-free medium if costs are a major concern.
16. TCA/acetone (27) may precipitate calcium-phosphate and small amino acids from the culture media. Very high contaminant concentrations may be achieved that extremely interfere with isoelectric focusing. Dialysis must not be used! Very high solute volumes are needed to remove salts and unspecific protein binding to the dialysis membrane may occur.
17. The 0.1 M PMSF stock solution is warmed up to 37°C in a water bath. PMSF is added before putting medium on ice to avoid precipitation. EDTA inhibits metalloproteases by chelation of free metal ions. PMSF inhibits serine proteases and some cysteine proteases. Inhibitor cocktails (for example Complete Mini, EDTA-free from Roche) must not be used! These cocktails contain protein- and peptide-based inhibitors that can reach very high concentrations upon ultrafiltration (up to 300X) and hence abundantly mask the secreted proteins present in the medium.
18. Usage guidelines for Centricon® Plus-70 centrifugal filter devices are given in the user guide (28). The filter material is made of a polyethersulfone Biomax membrane with a 5 kDa cut-off.
19. The pH of 9.0 from Tris serves two functions: 1) it maximizes protein extraction at basic pH conditions (almost any protein is in deprotonated state); 2) minimizes protease activity. The Tris itself serves as a positively charged ion that helps in solubilization of proteins (*see* Note 6). Very basic proteins may be lost!
20. Upon concentration protein precipitation may occur!
21. When Reagent B is first added to Reagent A, turbidity is observed that quickly disappears upon mixing to yield a clear, green colour.
22. Alternatively, wavelengths from 540-590 nm may be used with this method (34).

23. Amount of albumin standards (*see* Table 1) used: F, 0.625 µg; E, 3.125 µg; D, 6.25 µg; C, 12.5 µg; B, 18.75 µg and A, 25 µg. The standard amount is referred to a volume of 25 µL. Average blank-corrected 550 nm absorbance values above 0.8 must not be used for standard curve preparation!
24. For example: (protein amount of unknown is 7 µg) / (volume of diluted sample is 25 µL) x (dilution factor is 20)= 5.6 mg/mL.
25. A final volume of 450 µL is recommended by the supplier (Amersham Biosciences) for rehydration of one 24 cm Immobiline™ DryStrip. For first trial prepare a duplicate! The upper limit is twelve samples per run.
26. Avoid trapping of air bubbles.
27. Do not wear gloves during removal of protective cover sheet! The rubber material tends to stick to the „naked" gel and hence will damage it.
28. Overlaying of IPG strips with paraffin oil reduces risk of urea crystallization during rehydration.
29. In this case the paraffin oil evenly distributes the heat produced during IEF between the tray and the cooling plate.
30. Do not use ddH$_2$O or tap water! The former leads to very low conductivity between electrode and IPG strip, the latter to very high.
31. Each electrode has a side marked red or black.
32. Program for 24 cm IPG pH 3-10 NL strips using the EPS 3501 XL power supply (adapted from Hoving (39)): Phase 1, 300 V, 1 Vh (0.006 h); Phase 2, 300 V, 900 Vh (3 h); Phase 3, 3500 V, 9500 Vh (5 h) and Phase 4, 3500 V, 52500 Vh (15 h). Current and power are set non-limiting (2 mA, 5 W). Phases 1 to 4 are programmed in the gradient mode. The voltage will be ramping up to the maximum set in the phase, starting from zero in the first phase and in phases to follow from the end point of the phase before. Therefore in phases 1 and 3 voltage is linearly increased and in phases 2 and 4 held constant. The current check option must be switched off!

## 2.6 References

1. O'Farrell, P. H. (1975) High resolution two-dimensional electrophoresis of proteins. *J. Biol. Chem.* 250, 4007-4021.
2. Klose, J. (1975) Protein mapping by combined isoelectric focusing and electrophoresis in mouse tissues. A novel approach to testing for induced point mutations in mammals. *Humangenetik* 26, 231-243.
3. Lämmli, U. K. (1970) Cleavage of structural proteins during the assembly of the head of bacteriophage T4. *Nature* 227, 680-685.
4. O'Farrell, P. Z., Goodman, H. M., O'Farrell, P. H. (1970) High-resolution two-dimensional electrophoresis of basic as well as acidic proteins. *Cell* 12, 1133-1142.
5. Bjellqvist, B., Ek, K., Righetti, P. G., Gianazza, E., et al. (1982) Isoelectric focusing in immobilized pH gradients: principle, methodology and some applications. *J. Biochem. Biophys. Methods* 6, 317-339.
6. Görg, A., Postel, W., Günther, S. (1988) The current state of two-dimensional electrophoresis with immobilized pH gradients. *Electrophoresis* 9, 531-546.
7. Görg, A., Obermaier, C., Boguth, G., Harder, A., et al. (2000) The current state of two-dimensional electrophoresis with immobilized pH gradients. *Electrophoresis* 21, 1037-1053.
8. Görg, A., Weiss, W., Dunn, M. J. (2004) Current two-dimensional electrophoresis technology for proteomics. *Proteomics* 4, 3665-3685.
9. Matsudaira, P. (1987) Sequence from picomole quantities of proteins electroblotted onto polyvinylidene difluoride membranes. *J. Biol. Chem.* 262, 10035-10038.
10. Aebersold, R. H., Leavitt, J., Saavedra, R. A., Hood, L. E., Kent, S. B. (1987) Internal amino acid sequence analysis of proteins separated by one- or two-dimensional gel electrophoresis after in situ protease digestion on nitrocellulose. *Proc. Natl. Acad. Sci.* 84, 6970-6974.
11. Rosenfeld, J., Capdevielle, J., Guillemot, J. C., Ferrara, P. (1992) In-gel digestion of proteins for internal sequence analysis after one- or two-dimensional gel electrophoresis. *Anal. Biochem.* 203, 173-179.
12. Yates 3rd, J. R., Speicher, S., Griffin, P. R., Hunkapiller, T. (1993) Peptide mass maps: a highly informative approach to protein identification. *Anal. Biochem.* 214, 397-408.
13. James, P., Quadroni, M., Carafoli, E., Gonnet, G. (1994) Protein identification in DNA databases by peptide mass fingerprinting. *Protein Sci.* 3, 1347-1350.
14. Cottrell, J. S. (1994) Protein identification by peptide mass fingerprinting. *Pept. Res.* 7, 115-124.
15. Herbert, B. R., Sanchez, J.C., Bini, L. (1997) Two-dimensional electrophoresis: The state of the art and future directions, in *Proteome Research: New Frontiers in Functional Genomics* (Wilkins, M. R., Williams, K. L., Appel, R. D., Hochstrasser, D. F., eds.), Springer, Berlin, pp. 13-33.
16. Rabilloud, T., Adessi, C., Giraudel, A., Lunardi, J. (1997) Improvement of the solubilization of proteins in two-dimensional electrophoresis with immobilized pH gradients. *Electrophoresis* 18, 307-316.
17. Link, A. J. (ed.) (1999) *2-D Proteome Analysis Protocols*. Humana, Totowa, NJ.
18. Westermeier, R. (2001) *Electrophoresis in Practice*, 3rd Edition, Wiley-VCH, Weinheim.
19. Yates 3rd, J. R., Gilchrist, A., Howell, K. E., Bergeron, J. J. (2005) Proteomics of organelles and large cellular structures. *Nature* 6, 702-714.

20. Lim, J. W. E., Bodnar, A. (2002) Proteome analysis of conditioned medium from mouse embryonic fibroblast feeder layers which support the growth of human embryonic stem cells. *Proteomics* 2, 1187-1203.
21. Boraldi, F., Bini, L., Liberatory, S., Armini, A., et al. (2003) Normal human dermal fibroblasts: Proteomic analysis of cell layer and culture medium. *Electrophoresis* 24, 1292-1310.
22. Mears, R., Craven, R. A., Hanrahan, S., Totty, N., et al. (2004) Proteomic analysis of melanoma-derived exosomes by two-dimensional polyacrylamide gel electrophoresis and mass spectrometry. *Proteomics* 4, 4019-4031.
23. Prowse, A. B. J., McQuade, L. R., Bryant, K. J., Van Dyk, D. D., et al. (2005) A proteome analysis of conditioned media from human neonatal fibroblasts used in the maintenance of human embryonic stem cells. *Proteomics* 5, 978-989.
24. Volmer, M. W., Stühler, K., Zapatka, M., Schöneck, A., et al. (2005) Differential proteome analysis of conditioned media to detect Smad4 regulated secreted biomarkers in colon cancer. *Proteomics* 5, 2587-2601.
25. Eagle, H. (1959) Amino acid metabolism in mammalian cell cultures. *Science* 130, 432-437.
26. Richardson, J. C., Scalera, V., Simmons, N. L. (1981) Identification of two strains of MDCK cells which resemble separate nephron tubule segments. *Biochim. Biophys. Acta* 673, 26-36.
27. Damerval, C., DeVienne, D., Zivy, M., Thiellement, H. (1986) Technical improvements in two-dimensional electrophoresis increase the level of genetic variation detected in wheat-seedling protein. *Electrophoresis* 7, 53-54.
28. http://www.millipore.com/userguides.nsf/dda0cb48c91c0fb6852567430063b5d6/603b133b9b2a919c8525 6b3e0050b862/$FILE/P36006.pdf (User guide for Centricon® Plus-70 centrifugal filter devices from Millipore)
29. Bradford, M. M. (1976) A rapid and sensitive method for the quantitation of microgram quantities of protein utilizing the principle of protein-dye binding. *Anal. Biochem.* 72, 248-254.
30. Ramagli, L. S., Rodriguez, L. V. (1985) Quantitation of microgram amounts of protein in two-dimensional polyacrylamide gel electrophoresis sample buffer. *Electrophoresis* 6, 559-563.
31. Lowry, O. H., Rosebrough, N. J., Farr, A. L., Randall, R. J. (1951) Protein measurement with the Folin phenol reagent. *J. Biol. Chem.* 193, 265-295.
32. Mokrasch, L. C., McGilvery, R. W. (1956) Purification and properties of fructose-1, 6-diphosphatase. *J. Biol. Chem.* 221, 909-917.
33. Smith, R. K., Krohn, R. I., Hermanson, G. T., Mallia, A. K., et al. (1985) Measurement of protein using bicinchoninic adic. *Anal. Biochem.* 150, 76-85.
34. http://www.piercenet.com/files/1296dh4.pdf (Instructions for BCA™ Protein Assay Kit from Pierce)
35. Langen, H., Roder, D., Juranville, J. F., Fountoulakis, M. (1997) Effect of protein application mode and acrylamide concentration on the resolution of protein spots separated by two-dimensional gel electrophoresis. *Electrophoresis* 18, 2085-2090.
36. Rabilloud, T., Valette, C., Lawrence, J. J. (1994) Sample application by in-gel rehydration improves the resolution of two-dimensional electrophoresis with immobilized pH gradients in the first dimension. *Electrophoresis* 15, 1552-1558.

37. Sanchez, J. C., Rouge, V., Pisteur, M., Ravier, F., et al. (1997) Improved and simplified in-gel sample application using reswelling of dry immobilized pH gradients. *Electrophoresis* 18, 324-327.
38. Hoving, S., Voshol, H., van Oostrum, J. (2000) Towards high perfomance two-dimensional gel electrophoresis using ultrazoom gels. *Electrophoresis* 21, 2617-2621.
39. Hoving, S., Gerrits, B., Voshol, H., Muller, D., et al. (2002) Preparative two-dimensional gel electrophoresis at alkaline pH using narrow range immobilized pH gradients. *Proteomics* 2, 127-134.
40. http://www1.amershambiosciences.com/applic/upp00738.nsf/vLookupDoc/319798244-C534/$file/80642960.pdf (Amersham 2-D electrophoresis handbook)

# Chapter 3

## 3. Meprin – a metalloendopeptidase and its substrate repertoire – a degradomic approach[2]

### 3.1 Summary

In the past, protease-substrate discovery was rather haphazard and executed by *in vitro* cleavage assays using singly selected targets. Nowadays, with the rapidly growing field of proteomics, a protease and its substrate repertoire may be studied in complex biological systems; integrating information of a whole system's response to a protease's action. Here, we report the first proteomic approach applied to meprin, an astacin-like metalloendopeptidase, to determine physiologic substrates in a whole cell system. Human meprin$\alpha/\beta$ expressed on the cell surface of Madin-Darby canine kidney cells was activated by limited trypsin treatment. Culture media of these cells were subjected to two-dimensional gel electrophoresis, image analysis and liquid chromatography-based tandem mass spectrometry to identify proteins cleaved by meprin. Of 33 protein spots unique to media of cells expressing active cell surface meprin$\alpha/\beta$ 22 proteins were identified, and in follow-up experiments using immunoblotting the identity of proteins, namely, vinculin, lysyl oxidase, collagen type V and annexin A1 was confirmed. Upon classification into specific functional categories, ten proteins identified are associated with "cell growth and/or maintenance" and four with "immune response". Our findings therefore suggest a role for meprins in the regulation of cell homeostasis and the extracellular environment, and in innate immunity, respectively. Moreover, the identification of membrane proteins such as vesicular integral-membrane protein VIP36 in the culture medium indicates that meprin may be involved in ectodomain shedding.

---

[2] This book chapter has been published recently in : "Ambort D, Stalder D, Lottaz D, Huguenin M, Oneda B, Heller M and Sterchi EE (2008) A novel 2D-based approach to the discovery of candidate substrates for the metalloendopeptidase meprin. *FEBS J* **275**, 4490-4509."

## 3.2 Introduction

The astacin-like zinc-dependent metalloendopeptidase hmeprin was first discovered in 1982 due to its hydrolyzing activity towards the PABA-peptide (1). PABA-peptide hydrolase was subsequently purified and characterized from human small intestinal mucosa (2). At the same time PABA-peptide hydrolase orthologs, called meprin (metal endopeptidase from renal tissue) or endopeptidase-2, were found in mouse and rat kidney (3, 4). Hmeprin cDNA was cloned, sequenced and expressed in MDCK cells. Two similar subunits were identified and termed hmeprinα and hmeprinβ with molecular masses of 95 and 105 kDa, respectively. Hmeprinα was secreted into the culture medium of MDCK cells as inactive homodimers whereas hmeprinβ was membrane-bound (5). Hence heterodimers of hmeprinα/β allowed for localization of the α-subunit to the plasma membrane (6). Inactive zymogens of hmeprinα and β are processed by limited proteolysis with trypsin into their active forms (5, 6). Hmeprinα but not β may alternatively be activated by plasmin (7, 8).

A first step towards the elucidation of the biological function of meprin was achieved with testing of putatively cleavable polypeptide substrates. A variety of protein and peptide substrates were processed *in vitro*; biologically active peptides such as bradykinin, angiotensins I and II (2), polypeptide hormones, insulin B-chain and parathyroid hormone (9), as well as gastrointestinal peptides, namely, gastrin-releasing peptide fragment 14-27 and gastrin 17 (10), ECM components, collagen type IV, fibronectin and laminin-nidogen (11, 12), cytokines such as osteopontin (10), interleukin-1β and the chemokine monocyte chemoattractant protein-1 (13, 14). This suggests that meprin may be involved in the clearance of vasoactive peptides and polypeptide hormones from blood plasma, in the regulation of cell movement, the secretory activity and growth of intestinal tract and pancreas, in tissue remodelling processes, and finally, in the innate immune response. In addition, marked differences between α- and β-subunits in substrate and peptide bond specificity were reported and may point to distinct functions for the two forms (10). Meprinα selects for small (e.g. serine, alanine and threonine) or hydrophobic (e.g. phenylalanine) residues in P1 and P1' sites and proline in P2' position. Meprinβ prefers acidic amino acids in the P1 and P1' sites and selects against basic residues at P2' and P3'. In conclusion, protease-substrate discovery executed by these *in vitro* cleavage assays was rather haphazard. Thus meprin and its substrate repertoire may be studied in a complex biological context to identify physiologically relevant substrates.

The introduction of degradomics enabled identification of protease and protease-substrate repertoires on an organism-wide scale by means of proteomic techniques (15). A variety of hitherto unkown substrates for the metzincin metalloendopeptidases ADAM-17 and MMP-14 were found in conditioned media using different cell-based systems (16-18). Alternatively, a complex protein mixture, namely, human plasma, was digested with recombinant MMP-14 in a cell-free system (19). Two methodolocial platforms were successfully applied for protein separation; LC-MS/MS and 2-DE (16-19). These standard techniques were used in combination with lectin-affinity pre-fractionation and quantitative tags, namely, ICAT or cyanine dyes (DIGE). From these degradomic studies it became obvious that metalloendopeptidases are key modulators of diverse signalling pathways and not ECM destroying entities (20). For example, the major roles for the MMP family are the control of cellular responses critical to homeostatic regulation of the extracellular environment and the immune response (21, 22).

Hence we decided to apply degradomics in order to find novel physiologic substrates for meprin and to elucidate key functions on a system-wide level. For the above described techniques some conceptual problems may arise: ICAT-based approaches compare pairs of peptides, and therefore it is not possible to discover cleaved protein fragments with neo N- or C-termini. Furthermore, lectin-affinity purification with wheat germ agglutinin selects for oligosaccharides containing sialic acid or terminal N-acetylglucosamine. Consequently, N-glycosylated, O-glycosylated or non-glycosylated proteins not bearing those ligands may escape from such a pre-fractionation procedure. We thus designed a simple proteomic strategy that remedied those limitations and circumvented complicated quantitative and statistical evaluation. Hmeprinα/β was transfected into MDCK cells and activated *in situ* by limited trypsin treatment at confluent cell stage. Conditioned media of treated and non-treated cells were then concentrated with ultrafiltration. The salt-depleted protein samples were then separated by 2-DE. 2-D gel image analysis allowed for qualitative differential display of unique protein spots present in only one condition. Proteins of interest were successfully identified by LC-MS/MS and functionally classified.

## 3.3 Experimental procedures

### 3.3.1 Cell culture and meprin activation by in situ trypsin treatment

Wild-type and meprin$\alpha$/$\beta$ MDCK cells were grown in Minimum Essential Medium with Earle's salts supplemented with 5% (v/v) FBS, 100 units/ml penicillin and 100 µg/ml streptomycin (6, 23). For serum-free conditions same medium composition was used without FBS. $1.15 \times 10^6$ cells were plated in 100-mm dishes and incubated for about three days at 37 °C in an atmosphere of 5% $CO_2$ until cells were confluent. For limited trypsin treatment cells were washed twice with 4 ml of serum-free medium (5). Cells were then treated with 40 µl of trypsin solution (1 mg/ml in 50 mM Tris-HCl pH 7.5) diluted in 4 ml of serum-free medium for 30 min at 37 °C. For inactivation of trypsin cells were washed twice with 4 ml of serum-free medium. Cells were then incubated with 40 µl of soja bean trypsin inhibitor solution (2 mg/ml in water) diluted in 4 ml of serum-free medium for 30 min at 37 °C. Following inactivation cells were washed twice with 4 ml of serum-free medium. Negative controls were treated in the same way but without trypsin. Upon activation cells were conditioned in 4 ml of serum-free medium for 22 h at 37 °C.

### 3.3.2 Sample preparation of culture medium

Sample was prepared as previously described (24). The culture medium of 18 experimental replicates per condition (70-72 ml) was collected and defined as pooled biological replicate. Protease inhibitors (1 mM EDTA, 1 mM PMSF) were immediately added and the conditioned medium clarified by centrifugation for 1 h at $100,000 \times g$ at 4 °C in a fixed-angle rotor (TFT 70.38) on a KONTRON CENTRIKON T-2,060 ultracentrifuge (KONTRON Instruments AG, Zürich, Switzerland). Supernatants were concentrated 300-fold by ultrafiltration in Centricon® Plus-70 centrifugal filter devices (Millipore corporation, Billerica, MA, USA) at 4 °C. Concentrates were washed three times in sample solubilization buffer (20 mM Tris pH 9.0, 1 mM EDTA, 1 mM PMSF). Final protein concentration was determined with the BCA™ Protein Assay Kit (Pierce, Rockford, IL, USA).

### 3.3.3 One-dimensional gel electrophoresis

Three technical replicates (for image analysis) of one pooled biological replicate per conditioned medium sample were run through a polyacrylamide gel according to previously documented methods (25). Two technical gel replicates were produced for immunoblotting. A 12.5% T, 2.6% C resolving gel was first prepared and allowed to polymerize before overlaying with a 5% T, 2.6% C stacking gel. 10-30 µg protein was heated for 5 min in a reducing buffer (20 mM Tris-HCl pH 6.8, 10% (w/v) glycerol, 2% (w/v) SDS, 100 mM DTT) and then centrifuged for 5 min at 16,100 × g prior to loading. The proteins were electrophoresed at 5 mA/gel for 1h 30 min and 50 mA/gel for 50 min on a Mini-PROTEAN 3 Electrophoresis Cell (Bio-Rad Laboratories, Richmond, USA). 3 µl (for fluorescence staining) or 10 µl (for immunoblotting) of Precision Plus Protein™ All Blue Standards (Bio-Rad Laboratories, Richmond, USA) were loaded per gel.

### 3.3.4 Two-dimensional gel electrophoresis

2-DE was performed essentially as described (26). Three pooled biological replicates and two more technical replicates were run to have in total five 2-D gels per condition for subsequent analytical image analysis. For analytical gels 250 µg of concentrated medium protein from trypsin treated and non-treated MDCKα/β cells was solubilized in 450 µl of buffer containing 7 M urea, 2 M thiourea, 4% (w/v) CHAPS, 1% (w/v) DTE, 2% (v/v) Pharmalyte 3-10 for 1 h on a rotary shaker at room temperature. Sample-containing buffer was then centrifuged for 30 min at 16,100 × g before application to IPG strips (pH 3-10 NL, 24 cm, Amersham Biosciences, Uppsala, Sweden). Strips were rehydrated overnight in sample-containing buffer on the Immobiline™ DryStrip Reswelling Tray (Amersham Biosciences, Uppsala, Sweden) under paraffin oil. Focusing was always started at 300 V, and the voltage was slowly increased in a linear gradient to 3,500 V until a final volthour product of 63 kVh was reached. Focusing was performed on a Multiphor™ II horizontal electrophoresis apparatus (Amersham Biosciences, Uppsala, Sweden) under paraffin oil at 20 °C. After focusing the strips were equilibrated in 6 M urea, 30% (v/v) glycerol, 2% (w/v) SDS, 50 mM Tris-HCl pH 8.8 with 1% (w/v) DTE and 4.8% (w/v) iodoacetamide, respectively, each step for 15 min. For the second dimension strips were transferred to 12.5% T, 2.6% C SDS-PAGE gels (255 × 205 × 1.5 mm) using the Ettan™ Dalt*six* multiple vertical electrophoresis apparatus (Amersham Biosciences, Uppsala, Sweden) with following running conditions: 15 mA/gel for 1 h 30 min and 50 mA/gel for 5-6 h at 15 °C. For preparative gels 1 mg of protein was solubilized in 2 ml of buffer for 1 h at room temperature, centrifuged for 30 min at 16,100 × $g$, concentrated 10-fold in Centricon® YM-3 centrifugal filter devices (Millipore corporation, Billerica, MA, USA) and then diluted with buffer to a final volume of 450 µl. The total volthour product was increased to about 80 kVh. The cathodic paper wick was immersed in 1% (w/v) DTE before the run to counteract DTE depletion effects in the basic part of the strip.

### 3.3.5 Staining and imaging

Analytical 1-D or 2-D gels were visualized by post-electrophoretic fluorescence staining with ruthenium II tris (bathophenanthroline disulfonate). Ruthenium was synthesized exactly as described (27). Staining was performed according to the improved protocol (28). In addition to the standard procedure, gels were incubated in 20 mM Tris for 30 min at 80 rpm, washed twice in deionized water for 10 min and finally, destained again in 40% (v/v) ethanol, 10% (v/v) acetic acid overnight at 80 rpm. The next day gels were rinsed twice in deionized water for 10 min before scanning on a Fuji Film Fluorescent Image Analyzer FLA-3,000R with control software BASReader version 3.01 (Raytest Isotopenmessgeräte GmbH, Straubenhardt, Germany). Images were digitized using following parameters: 473 nm excitation, O580 filter, sensitivity 1,000, 16 bits per pixel, 50 µm pixel size. Images saved in Fuji BAS file format were converted to 16 bit per pixel Tagged Image File Format images with AIDA version 3.11 (Raytest Isotopenmessgeräte GmbH, Straubenhardt, Germany). Preparative 2-D gels were stained by colloidal Coomassie brilliant blue G-250 and used for subsequent protein identification (29).

### 3.3.6 Image analysis

1-D image analysis was performed using the 1-D Evaluation module of AIDA version 3.11. In total three technical gel replicates with pooled biological replicates (from 18 dishes) of each condition were produced. The pooled biological replicates of each condition were loaded on the same gel. Three independent quantitative analysis were done. Gel image attributes were defined in quantum levels and pixel (65,536 quantum levels per pixel). Rectangular densitometer windows (100 pixels in width over entire lane) were used to generate profile scans of each gel lane. In each profile scan vertical peak borders were defined to subdivide the whole gel lane into integrable major peaks. Following baseline correction the averaged signal intensity integrated over each defined peak was plotted against its number.

A 2-D gel image analysis strategy was designed to circumvent cumbersome quantitative comparison and statistical evaluation. The strategy is based on qualitative changes among master or reference gels (level 1 matchsets) of each group of five gel replicates. In total three pooled biological gel replicates (from 18 dishes per pooled sample) and two more technical gel replicates (of one representative pooled sample) were produced per condition. 2-D gel image analysis was performed using the program PDQuest version 7.3.1 (Bio-Rad Laboratories, Richmond, CA, USA). Data was inverted and images displayed as black spots on white background. Each gel replicate was cropped into four quadrants of same image size (2,157 × 1,682 pixels). Image cropping of same areas among gel replicates was realized by reference to highly conserved landmark spots present in all gels. In total four independent analysis were done on each of the four quadrant sections. Therefore corresponding quadrant sections of gel replicates from the same condition were grouped into level 1 matchsets. Gel sections were filtered (Median, 9 x 9 pixels) to remove image noise. Spots were detected as follows: sensitivity 20.00, size scale 9, minimum peak 4,230. Background was subtracted with floater method (radius size, 67 pixels). Spot editing was done only within same condition members. Spot matching was done within same condition members of the same level 1 matchset. In order to find qualitative changes between different conditions the level 1 matchsets were clustered into super level 1 matchsets (higher level matchsets). Matching was then performed on the master gels of the level 1 matchsets. All higher level matchsets were combined into one super higher level matchset (combined higher level matchset). Qualitative changes could then be displayed as sets of unique spots present in only one or the other condition.

## *3.3.7 Protein identification by liquid chromatography-based tandem mass spectrometry and protein database searching*

Gel plugs containing protein spots of interest were excised and the proteins were subjected to in-gel tryptic digestion and peptide extraction as described (30). 20 µl of protein digest was loaded onto a self-made microbore column (0.15 mm i.d. × 80 mm length) at a flow rate of ~4 µl/min of solvent A (0.1% (v/v) formic acid in water/ACN 98:2). Columns were packed with GROM-SIL 300 Octyl-6 MB, 5 mm, reversed-phase material (Grom GmbH, Rottenburg-

Haiflingen, Germany). Columns were developed by a bi-phasic ACN gradient of 0 to 5% solvent B (0.1% (v/v) formic acid in water/ACN 4.9:95) in 1 min followed by 5 to 40% solvent B in 20 min at a flow rate of approximately 3 ml/min. The column effluent was directly coupled to an Esquire3,000+ ion trap mass spectrometer from Bruker Daltonics (Bremen, Germany) via a capillary ESI source operated at 3,700 V. CID was triggered on the two most abundant not singly charged peptide ions in the $m/z$ range of 360-1,400. Precursors were set in an exclusion list for 0.5 min. Peak lists from the raw data were created by Data Analysis version 3.1 (Bruker Daltonics, Bremen) using the following parameters. MS/MS compounds exceeding a total ion chromatogram intensity of 4000 ion counts were exported and all spectra from the same precursor eluting within a retention time window of 0.5 minutes were compiled to one MS/MS peak list. MS/MS peak detection was made with the Apex peak finder algorithm using a peak width at half maximum (FWHM) of 0.1 $m/z$, a signal-to-noise ratio (S/N) of one, a relative to base peak intensity threshold of 2%, and an absolute intensity threshold of 10 ion counts as parameters. A mixed list of deconvoluted and non-deconvoluted MS and MS/MS signals, with an allowance for only the 200 most abundant peaks from non-deconvoluted MS/MS signals of each spectrum, were exported into Mascot generic file format text (mgf) files. MS signal deconvolution was set to "Auto" for resolved isotope, and a maximum charge of four with minimally three peaks in set and a molecular weight agreement of 0.05% for related ion deconvolution, respectively. MS/MS peak deconvolutions were allowed for a maximum charge of one only. S/N and FWHM values were also exported into the mgf files. CID spectra interpretation was performed with the public search engine Phenyx version 2.1 on vital-it.ch server operated by GeneBio (Geneva, Switzerland) against the uniprot-SwissProt protein database (release 48.8) with fixed carbamidomethyl modification of cysteine residues, variable oxidation of methionine and variable deamidation of asparagine and glutamine. Parent and fragment mass tolerances were set to 1 Da. Up to two missed cleavages and half tryptic peptides were allowed. The taxonomic search space was restricted to Mammalia (40,084 sequence entries). Peptide search criteria were set to a minimum peptide z-score of ≥5 and a maximum peptide p-value of ≤0.0001. All protein identifications consisting of at least two unique peptides reaching a p-value of ≤0.00000001 were accepted. To double-check significant hits same spectra were interpreted with the web-based search engine MASCOT version 2.1 operated by Matrix Science Ltd. (London, UK) against same database and parameter settings as above (31). To identify proteins not previously described for dog all significant peptide matches were searched with program BLASTP version 2.2.16 against the dog genome database (32, 33). Database size was 33,527 dog RefSeq protein sequences. This database is hosted at NCBI (Bethesda, MD, USA). Searches were performed as follows: word size 3, filter low complexity, expect value 0.01, score matrix BLOSUM62. Failed searches were repeated with settings for "short and nearly exact matches": word size 2, filter off, expect value 20,000, score matrix PAM30. Only the top scoring significant hit was accepted.

### *3.3.8 Functional classification*

Proteins were clustered into functional groups according to the Human Protein Reference Database (http://hprd.org/) (34). The corresponding human orthologs were grouped and subgrouped into biological process and molecular function.

### *3.3.9 Immunoblotting*

Western blotting was performed as described (35). Following one-dimensional gel electrophoresis as detailed above the protein was transferred to a Hybond™-P PVDF membrane (Amersham Biosciences, Uppsala, Sweden) by application of a constant potential for 15 min at 30 V and 50 min at 80 V. The membrane was then incubated overnight at 4 °C in TTBS (20 mM Tris-HCl pH 7.5, 137 mM NaCl, 0.1% (w/v) Tween-20) containing 5% (w/v) milk powder. The membrane was washed twice in TTBS for 1 min and 10 min and incubated for 1 h with primary antibody prepared in TTBS containing 2% (w/v) milk powder. Thereafter the membrane was washed four times in TTBS. The secondary antibody was horseradish peroxidase-linked donkey anti-rabbit or sheep anti-mouse (Amersham Biosciences, Uppsala, Sweden) diluted 1:10,000 in antibody solution. The membrane was incubated for 1 h and washed four times in TTBS. Immunoblots were analyzed using the ECL plus Western Blotting Detection System (Amersham Biosciences, Uppsala, Sweden). Monoclonal antibodies against vinculin (diluted 1:2,000) and annexin A1 (diluted 1:2,000) were a generous gift of Dr. Eduard B. Babiychuk (Department of Cell Biology, Institute of Anatomy, University of Berne, Switzerland). Polyclonal antibody against lysyl oxidase (diluted 1:4,000) was purchased from IMGENEX Corporation (San Diego, CA, USA). Monoclonal antibody 1E2-E4/Col5 against collagen type V (diluted 1:1,000) was from CHEMICON Australia Pty. Ltd. (Victoria, Australia) (36).

## 3.4 Results

### *3.4.1 Meprin activation by in situ trypsin treatment*

The key to reliable proteomic data is the establishment of a highly reproducible standardized sample preparation procedure and a robust experimental system. Zymogen activation by limited trypsin treatment may introduce disturbances into such a system. In order to exclude unspecific side effects caused by the trypsin treatment rather than by the effector (membrane-bound hmeprin$\alpha/\beta$), wild-type and meprin$\alpha/\beta$ MDCK cells were treated in the same way. Media of treated and non-treated cells were prepared and then subjected to one-dimensional gel electrophoresis and subsequent densitometric image analysis with AIDA software (Fig. 1). Visible appearance of major bands on 1-D gels showed no differences between treated and non-treated WT samples whereas meprin$\alpha/\beta$ samples showed substantial changes upon trypsin activation. Moreover, the 1-D patterns of WT versus meprin$\alpha/\beta$ differed as well (Fig. 1A) and indicated overexpression of hmeprin$\alpha/\beta$ per se causing changes that are independent from the stimulus. Quantitative image analysis of 1-D gel lanes confirmed these findings (Fig. 1B) but more importantly, revealed a trend towards the appearance of low molecular weight proteins in media of meprin$\alpha/\beta$ MDCK cells. Repetition of experiments unambiguously pointed to reproducible changes triggered by the activation and not by the overexpression of meprin$\alpha/\beta$ (Fig. 1C). Obviously, activation of meprin$\alpha/\beta$ results in the release of proteins into the culture medium. Thus we hypothesized that upon trypsin activation of meprin$\alpha/\beta$ on the surface of MDCK cells, additional proteins may be present in the culture medium that may be separated and identified by means of proteomic techniques.

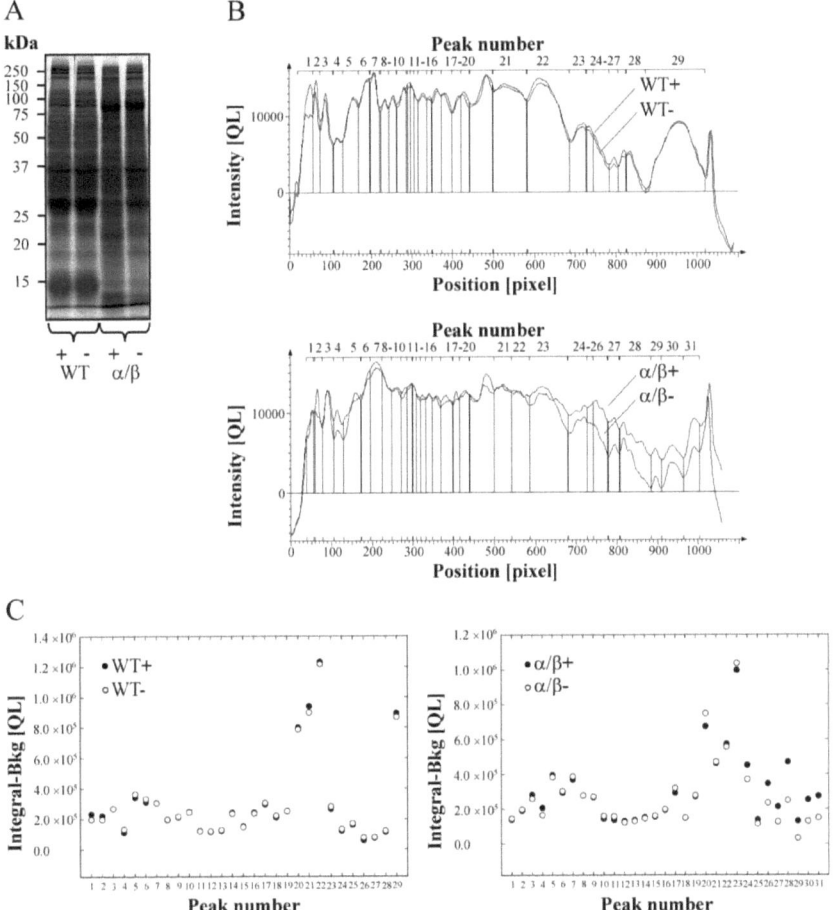

**Fig. 1. Image analysis of 1-D gels.**

(A) Representative 1-D separation of conditioned medium protein (20 μg per lane) from trypsin treated (+) and non-treated (-) wild-type and meprinα/β MDCK cells in a 12.5% SDS gel under reducing conditions. Optimized Ruthenium staining. Migration positions of molecular mass standards are shown on the gel. In total three independent technical gel replicates were produced. (B) Densitometric analysis of profile scans from a representative 1-D gel. For each lane a rectangular densitometric window was used to graphically display pixel intensity (quantum levels, QL) versus migration position (pixel). Peaks were subdivided into integrable areas and numbered. WT (upper graph) and meprinα/β profiles (lower graph) were then superimposed. (C) Averaged quantitative comparison of WT (left hand side) and meprinα/β peaks (right hand side) from three independent analysis. Intensity of peak areas (QL) was background-corrected (Bkg).

## *3.4.2 Design and application of a novel two-dimensional gel image analysis strategy*

Traditionally, 2-D gel image analysis is performed on two sets of gels and protein spots are matched to the same master or reference gel within one analysis. Statistical tools (Student's t-test, Wilcoxon test) may then quantitatively assess subtle but significant changes in peak volumes to find up- or down-regulated protein spots. Unfortunately, error-prone matching to wrong reference spots is often underestimated overwhelming the benefits of quantitative statistical information. A remedy to false-positive data interpretation is the stepwise reduction in complexity of such analysis. Therefore we designed a novel image analysis strategy in which digitized 2-D gels were cut into four quadrants to maximize operator's overview. The corresponding quadrant sections were modularly grouped into sets of gels (level 1 matchsets) for each condition and then into supersets of level 1 matchsets (higher level matchsets) (Fig. 2). This strategy allowed for subsequent matching of protein spots first to reference gels of the same condition and thereafter to reference gels common to both conditions. The modular character of this type of analysis allowed for differential display of spots unique to each condition and quadrant section without relying on any quantitative information (Fig. 3).

Applying the above strategy to conditioned media of MDCKα/β cells revealed that among 817 protein spots displayed, 35 (4.3%) were unique to media of cells expressing active meprinα/β and 40 (4.9%) to media of cells with non-activated meprinα/β (Table 1). Thus unique spots were indicative of proteins released into or proteolytically cleaved in the extracellular milieu by hmeprinα/β. The compatibility of our sample preparation method and image analysis strategy to classical DIGE was further verified using differential labelling of media concentrates with Cy3 and Cy5. Of 1,197 protein spots considered in total, 79 (6.6%) decreased and 61 (5.1%) increased by a factor of 2.5 in maximum spot volume upon meprinα/β activation (data not shown). For simplification of work we decided to identify only proteins released into culture medium upon activation of meprinα/β. Protein identification was performed by LC-MS/MS as detailed below.

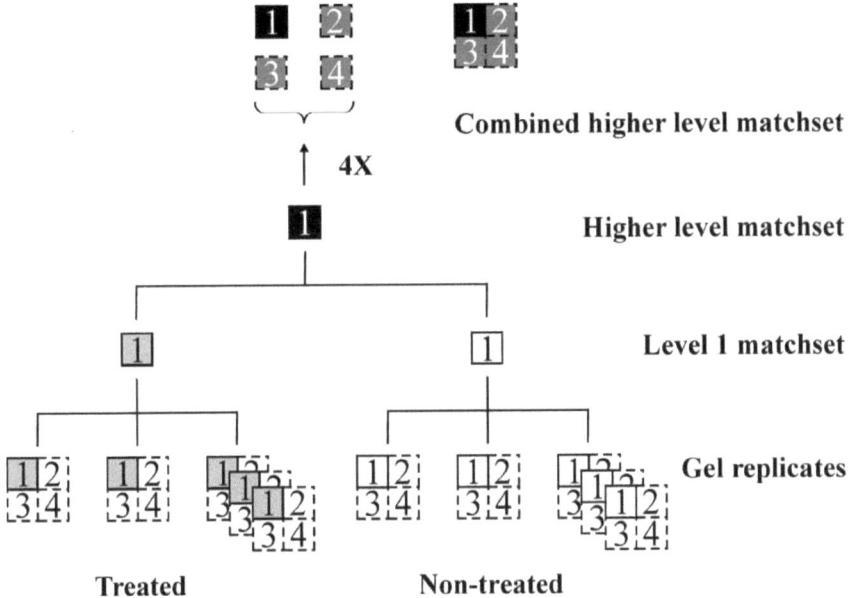

**Fig. 2. Image analysis strategy for two-dimensional gels.**
The strategy is based on qualitative changes among reference gels (level 1 matchsets) of each group of five gel replicates (three pooled biological gel replicates and two more technical gel replicates). Gel replicates of each group (treated versus non-treated) were cut virtually into four equally spaced quadrant sections to undergo independent image analysis. Reference gels of each group were then clustered into a new matchset for higher level differential display. The spot matching features of PDQuest version 7.3.1 allowed for visualization of unique protein spots present in one group or vice versa. The combined higher level matchset is the final fusion of all annotated unique spots into one big reference map.

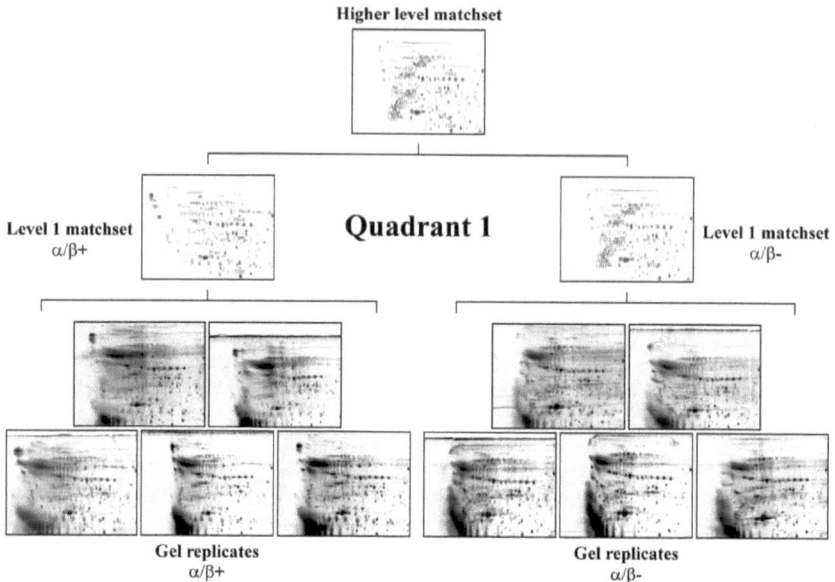

**Fig. 3. Quadrant section-based image analysis of two-dimensional gels.**

A representative image analysis of the first quadrant section is shown. 250 μg of conditioned medium protein from trypsin treated (+) and non-treated (-) MDCKα/β cells was separated by IEF in a 24 cm long IPG pH 3-10 NL strip. Vertical separation according to mass in a 12.5% SDS gel. Optimized Ruthenium staining. For each condition three pooled biological gel replicates (from 18 dishes per pooled sample) and two more technical gel replicates (of one pooled sample) were produced for subsequent image analysis. Unique protein spots are labelled in level 1 and higher level matchsets with SSP assigned by the image analysis software.

**Table 1**

**Protein spot matching statistics**

2-D gel image analysis was performed with PDQuest version 7.3.1 on five gel replicates (three biological replicates, two technical replicates) of conditioned media from trypsin treated and non-treated MDCKα/β cells. Qualitative spot matching differences among reference gels (level 1 matchsets) are expressed as unique spots (in percentage of each corresponding quadrant section).

| Condition | Quadrant | Gel replicates | | | | | Master | | Unique spots (in %) |
|---|---|---|---|---|---|---|---|---|---|
| | | Replicate 1 | Replicate 2 | Replicate 3 | Replicate 4 | Replicate 5 | Level 1 matchset | Higher level matchset | |
| Trypsin treated | 1 | 315 | 313 | 318 | 315 | 316 | 318 | | 2 (0.6) |
| Non-treated | 1 | 334 | 333 | 334 | 332 | 332 | 334 | | 18 (5.4) |
| In Total | 1 | | | | | | | 336 | |
| Trypsin treated | 2 | 221 | 219 | 221 | 215 | 218 | 222 | | 10 (4.4) |
| Non-treated | 2 | 217 | 212 | 218 | 216 | 217 | 218 | | 6 (2.6) |
| In Total | 2 | | | | | | | 228 | |
| Trypsin treated | 3 | 106 | 103 | 116 | 115 | 117 | 122 | | 12 (9.2) |
| Non-treated | 3 | 107 | 110 | 113 | 103 | 104 | 119 | | 9 (6.9) |
| In Total | 3 | | | | | | | 131 | |
| Trypsin treated | 4 | 105 | 107 | 115 | 108 | 105 | 115 | | 11 (9.0) |
| Non-treated | 4 | 110 | 108 | 109 | 102 | 104 | 111 | | 7 (5.7) |
| In Total | 4 | | | | | | | 122 | |
| Trypsin treated | all | | | | | | 777 | | 35 (4.3) |
| Non-treated | all | | | | | | 782 | | 40 (4.9) |
| In Total | all | | | | | | | 817 | |

## 3.4.3 Identification of dog protein orthologs by means of liquid chromatography-based tandem mass spectrometry, Phenyx-based and BLASTP-based protein database searching

By visual inspection the 35 protein spots unique to media of trypsin treated MDCKα/β cells could be reduced to 33 putative candidates. The redundancy of two spots present in more than one quadrant section-based image analysis prompted correction (Fig. 3 and supplementary Fig. S1-S3). On colloidal Coomassie stained preparative 2-D gels 24 (73%) protein spots of interest were detectable. These spots could be rematched to putative candidates found in fluorescence stained analytical gels (data not shown). Gel plugs were then prepared, digested in-gel with trypsin and peptides thereof separated/fragmented by LC-MS/MS. CID spectra interpretation with Phenyx version 2.1 against uniprot-SwissProt protein database (release 48.8) led to 22 (67%) protein identifications (Fig. 4 and Table 2). The taxonomic search space was restricted to Mammalia (40,084 sequence entries). To double-check significant hits the same spectra were interpreted with the web-based search engine MASCOT version 2.1 against the same database and parameter settings (data not shown) (31). The identification of nucleophosmin (protein spot SSP 2102, Table 2) was accepted because the peptide VDNDENEHQLSR and its in-source produced fragment DNDENEHQLSLR were unambiguously identified with good scores by Phenyx and Mascot. In addition, the fully tryptic peptide MSVQPTVSLGGFEITPPVVLR was identified by Phenyx and Mascot as first ranking identification, however with scores below the chosen acceptance criteria (Table 2 and data not shown). Beside six (27%) positive hits for dog, other species such as rat, human, rabbit and mouse were predominantly represented. The current release (51.3) of uniprot-SwissProt protein database lists 664 sequence entries for dog and thus may explain the poor representation in this species. Recently, the dog genome was sequenced to completion (33). Peptide sequence tags deciphered from our previous analysis permitted search with BLASTP version 2.2.16 against NCBI's 33,527 dog RefSeq protein sequence entries (32). Finally, all equivocal uniprot-SwissProt protein database searches were successfully matched to predicted dog protein orthologs (Table 3).

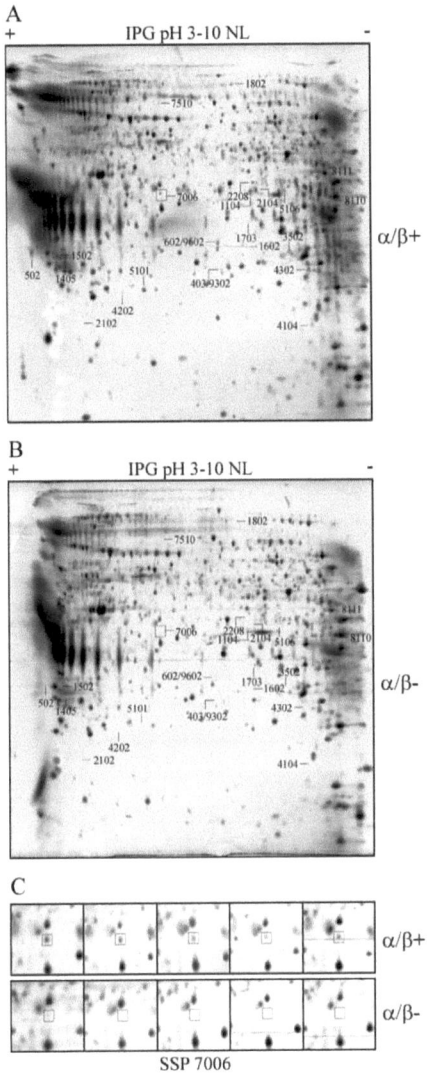

**Fig. 4. Two-dimensional reference maps on protein identifications.**

Representative 2-D gel images of conditioned medium protein from MDCKα/β cells. (A) 2-D gel of trypsin treated (+) condition. (B) 2-D gel of non-treated (-) condition. Protein spots unique to trypsin treated condition were labelled with SSP defined by image analysis software. (C) Close-up view of one representative protein spot, namely, SSP 7006, present in all gel replicates of trypsin treated condition. Protein spots of interest were identified by in-gel tryptic digestion and LC-MS/MS (Table 2).

**Table 2**

**Protein identification by liquid chromatography-based tandem mass spectrometry**

In-gel tryptic digestion of 2-D gel protein spots unique to conditioned media of trypsin treated MDCKα/β cells. Peptide separation/fragmentation with LC-MS/MS. CID spectra interpretation with public search engine Phenyx version 2.1 on vital-it.ch against the uniprot-SwissProt protein database (release 48.8). Taxonomic search space restricted to Mammalia (40,084 sequence entries). SSP assigned by image analysis software PDQuest version 7.3.1. CANFA, *Canis familiaris*, dog; RAT, *Rattus norvegicus*, rat; HUMAN, *Homo sapiens*, human; RABIT, *Oryctolagus cuniculus*, rabbit; MOUSE, *Mus musculus*, mouse. For multiple peptide matches to same primary sequence the top scoring peptide was listed. Modifications: C^, carbamidomethylation of cysteine; M*, oxidation of methionine; Q#, deamidation of glutamine. Half-cleaved tryptic peptides are indicated (in italic): [a] N- and C- terminal half-cleaved peptides; [b] half-cleaved peptides generated during in-gel digestion, [c] in-source fragmentation products, [d] normal tryptic peptide (dog protein ortholog with arginine in P1 position instead of glutamine). Half-cleaved neo peptides are highlighted (in bold). Match delta is the difference between theoretical $m/z$ of matched peptide and observed $m/z$ of parent ion. Peptide search criteria were set to a minimum peptide z-score of ≥5 and a maximum peptide p-value of ≤0.0001. All protein identifications consisting of at least two unique peptides reaching a p-value of ≤0.00000001 were accepted.

| SSP | Protein identification | Swiss-Prot accession number | Number of unique peptides | Sequence | Experimental $m/z$ (Th) | Theoretical mass (Da) | Match delta $m/z$ (Th) | Peptide z-score | Peptide p-value |
|---|---|---|---|---|---|---|---|---|---|
| 502 | CLUS_CANFA | P25473 | 11 | [a]*-I/DQAVSDTELQEM*STEGSK* | 985.952 | 1969.842 | -0.023 | 9.81 | 6.45e-18 |
| | | | | ***SDTELQ#EM*STEGSK*** | 735.824 | 1470.603 | 0.485 | 9.25 | 1.72e-15 |
| | | | | K/TLIEQTNEER | 616.842 | 1231.604 | -0.032 | 11.3 | 1.82e-24 |
| | | | | K/TLIEQTNEERK | 680.894 | 1359.699 | -0.037 | 6.36 | 1.63e-05 |
| | | | | R/KSLLSNLEEAK | 616.363 | 1230.682 | -0.014 | 8.68 | 3.46e-13 |
| | | | | K/SLLSNLEEAKK | 616.421 | 1230.682 | -0.072 | 7.14 | 8.22e-08 |
| | | | | K/SLLSNLEEAK | 552.277 | 1102.587 | 0.024 | 11.7 | 1.79e-26 |
| | | | | K/EDALNDTKDSETK | 732.427 | 1464.658 | 0.91 | 8.96 | 5.08e-14 |
| | | | | R/IDSLLENDR | 537.749 | 1073.535 | 0.026 | 8.04 | 8.17e-11 |
| | | | | [b]*R/QQTTHALDVM*Q/D* | 593.764 | 1185.544 | 0.016 | 6.31 | 5.16e-05 |
| | | | | R/ASSIM*DELFQDR | 714.356 | 1426.639 | -0.029 | 7.59 | 2.53e-09 |
| 403/9302 | LYOX_RAT | P16636 | 4 | R/C^AAEENC^LASSAYR | 801.35 | 1600.661 | -0.012 | 12.9 | 3.00e-33 |
| | | | | K/ASFC^LEDTSC^DYGYHR | 661.002 | 1979.778 | -0.069 | 7.82 | 5.07e-10 |
| | | | | K/VSVNPSYLVPLSDYSNNVVR | 1119.092 | 2237.096 | 0.464 | 8.29 | 6.36e-12 |
| | | | | [a]*R/TGHHAYASGC^TSPY/-I* | 892.899 | 1783.762 | -0.01 | 6.27 | 2.29e-05 |
| 602/9602 | LMAN2_CANFA | P49256 | 12 | [a]*-I/DITDGNSEHLKR* | 692.894 | 1383.674 | -0.049 | 7.81 | 4.37e-10 |
| | | | | [c]*I/TDGNSEHLKR* | 578.845 | 1155.563 | -0.056 | 7.5 | 1.23e-08 |

58

| | | | | Peptide | | | | | |
|---|---|---|---|---|---|---|---|---|---|
| | | | | *T/DGNSEHLKR* | 528.275 | 1054.515 | -0.01 | 7.39 | 2.74e-08 |
| | | | | K/NLHGDGIALWYTR | 758.428 | 1514.763 | -0.039 | 12 | 2.25e-28 |
| | | | | R/LVPGPVFGSK | 500.22 | 999.575 | 0.575 | 6.61 | 7.88e-06 |
| | | | | K/DNFHGLAIFLDTYPNDETTER | 1235.101 | 2467.129 | -0.529 | 10.3 | 5.12e-20 |
| | | | | R/WTELAGC*TADFR | 713.858 | 1425.634 | -0.033 | 10.6 | 3.30e-21 |
| | | | | R/NRDHDTFLAVR | 448.599 | 1342.674 | -0.034 | 7.24 | 5.41e-08 |
| | | | | R/DHDTFLAVR | 537.316 | 1072.53 | -0.043 | 11.8 | 5.80e-27 |
| | | | | R/LTVM*TDLEDKNEWK | 869.483 | 1736.829 | -0.061 | 6.41 | 9.90e-06 |
| | | | | R/LPTGYYFGASAGTGDLSDNHDIISM*K | 916.184 | 2745.259 | -0.09 | 8.27 | 2.00e-11 |
| | | | | *KLFQLM*VEHT* | 516.781 | 1031.511 | -0.018 | 7.11 | 1.13e-07 |
| 1104 | CO5A2_HUMAN | P05997 | 2 | K/SLSSQIETM*R | 584.217 | 1166.56 | 0.071 | 10.7 | 1.43e-21 |
| | | | | R/GSQFAYGDHQSPNTAITQM*TFLR | 862.413 | 2585.197 | 0.327 | 7 | 3.60e-07 |
| 1405 | CLUS_CANFA | P25473 | 3 | K/TLIEQTNEER | 616.738 | 1231.604 | 0.072 | 9.69 | 2.75e-17 |
| | | | | K/SLLSNLEEAK | 552.23 | 1102.587 | 0.071 | 8.01 | 2.18e-10 |
| | | | | R/ASSIM*DELFQDR | 714.245 | 1426.639 | 0.082 | 10.7 | 4.56e-22 |
| 1502 | CLUS_CANFA | P25473 | 12 | *-L/DQAFSDTELQEM*STEGSK* | 985.976 | 1969.842 | -0.047 | 12.8 | 9.20e-33 |
| | | | | *SDTELQ#EM*STEGSK* | 735.838 | 1470.603 | 0.471 | 9.42 | 3.50e-16 |
| | | | | K/TLIEQTNEER | 616.818 | 1231.604 | -0.008 | 8.08 | 5.51e-11 |
| | | | | K/TLIEQTNEERK | 680.9 | 1359.699 | -0.043 | 6.85 | 6.10e-07 |
| | | | | K/SLLSNLEEAK | 552.847 | 1102.587 | -0.546 | 13.4 | 1.00e-35 |
| | | | | K/EDALNDTKDSETK | 733.334 | 1464.658 | 0.003 | 6.5 | 6.00e-06 |
| | | | | R/IDSLLENDR | 537.774 | 1073.535 | 0.001 | 9.28 | 1.59e-15 |
| | | | | [b]R/IDSLLENDRQQTHALI/D | 877.014 | 1751.88 | -0.066 | 7.09 | 9.27e-08 |
| | | | | [b]R/QQTHALDVM*Q/D | 593.776 | 1185.544 | 0.004 | 7.62 | 2.11e-09 |
| | | | | R/Q#QTHALDVM*QDSFNR | 903.468 | 1805.8 | 0.44 | 10.1 | 2.74e-19 |
| | | | | R/ASSIM*DELFQDR | 714.365 | 1426.639 | -0.038 | 8.97 | 4.82e-14 |
| | | | | K/LYDELLQSYQEK | 764.85 | 1527.745 | 0.03 | 7.62 | 3.90e-09 |
| 1602 | LMAN2_CANFA | P49256 | 9 | [a]-L/DITDGNSEHLKR | 692.816 | 1383.674 | 0.029 | 8.59 | 6.59e-13 |
| | | | | *L/TDGNSEHLKR* | 578.799 | 1155.563 | -0.01 | 6.43 | 1.04e-05 |
| | | | | K/NLHGDGIALWYTR | 757.873 | 1514.763 | 0.516 | 13.4 | 2.85e-36 |
| | | | | *V/PGPVFGSK* | 394.734 | 787.422 | -0.015 | 9.47 | 3.28e-16 |
| | | | | K/DNFHGLAIFLDTYPNDETTER | 1234.648 | 2467.129 | -0.076 | 9.1 | 4.37e-15 |
| | | | | R/WTELAGC*TADFR | 713.829 | 1425.634 | -0.004 | 7.65 | 3.25e-09 |

| | | | | | | | | | | |
|---|---|---|---|---|---|---|---|---|---|---|
| 1703 | ANXA1_RABIT | P51662 | | R/DHDTFLAVR | 537.151 | 1072.53 | 0.122 | 8.88 | 1.30e-13 |
| | | | | K/NC^IDITGVR | 524.223 | 1046.517 | 0.043 | 9.45 | 3.29e-16 |
| | | | | *K/LFQLM*V*EH/T* | 516.778 | 1031.511 | -0.015 | 7.94 | 4.22e-10 |
| 1802 | FLNA_MOUSE | Q8BTM8 | 2 | K/GVDEATIIDILTK | 694.331 | 1386.76 | 0.057 | 12.8 | 1.54e-32 |
| | | | | K/TPAQFDADELR | 631.73 | 1261.593 | 0.074 | 10.9 | 1.56e-22 |
| | | | 4 | K/DAGEGGLSLAIEGPSK | 750.424 | 1499.746 | 0.457 | 9.27 | 1.43e-15 |
| | | | | K/VDINTEDLEDGTC^R | 818.007 | 1635.704 | 0.853 | 8.84 | 1.46e-13 |
| | | | | R/EAGAGGLAIAVEGPSK | 713.941 | 1425.746 | -0.06 | 7.27 | 2.85e-08 |
| | | | | K/VNQPASFAVSLNGAK | 752.41 | 1501.788 | -0.508 | 10.9 | 5.55e-23 |
| 2102 | NPM_RAT | P13084 | 3 | K/VDNDENEHQLSLR | 784.81 | 1567.722 | 0.059 | 8.1 | 3.99e-11 |
| | | | | *V/DNDENEHQLSLR* | 735.34 | 1468.654 | -0.005 | 8.31 | 7.53e-12 |
| | | | | K/M*SVQPTVSLGGFEITPPVVLR | 1121.629 | 2242.203 | 0.48 | 6.52 | 7.52e-06 |
| 2104 | COS A2_HUMAN | P05997 | 2 | K/SLSSQIETM*R | 584.201 | 1166.56 | 0.087 | 10.9 | 1.26e-22 |
| | | | | R/GSQFAYGDHQSPNTAITQM*TFLR | 862.836 | 2585.197 | -0.096 | 7.27 | 4.86e-08 |
| 2208 | CAPG_HUMAN | P40121 | 2 | K/YQEGGVESAFHK | 676.259 | 1350.62 | 0.059 | 8.03 | 1.65e-10 |
| | | | | *Q/YAPNTQVEILPQGR* | 793.287 | 1584.826 | 0.134 | 11.9 | 6.76e-28 |
| 3502 | LOXL1_HUMAN | Q08397 | 2 | K/C^LASTAYAPEATDYDVR | 951.853 | 1901.846 | 0.078 | 9.85 | 8.28e-18 |
| | | | | K/YIVLESDFTNNVVR | 834.825 | 1667.851 | 0.108 | 14.1 | 2.75e-40 |
| 4104 | STC1_HUMAN | P52823 | 3 | R/M*IAEVQEEC*YSK | 751.869 | 1501.642 | -0.04 | 8.45 | 2.22e-12 |
| | | | | K/RNPEAITEVVQ#LPNHFSNR | 741.405 | 2221.124 | -0.023 | 10.1 | 1.04e-18 |
| | | | | R/SLLEC^DEDTVSTIR | 818.872 | 1636.761 | 0.516 | 10.1 | 2.41e-19 |
| 4202 | CLUS_CANFA | P25473 | 8 | *a_1/DQAI*SDTELQEM*STEGSK | 985.781 | 1969.842 | 0.148 | 7.09 | 8.33e-08 |
| | | | | K/TLIEQTNEER | 616.785 | 1231.604 | 0.025 | 10.8 | 5.30e-22 |
| | | | | K/TLIEQTNEERK | 680.757 | 1359.699 | 0.1 | 8.63 | 4.85e-13 |
| | | | | K/SLLSNLEEAK | 552.235 | 1102.587 | 0.066 | 8.81 | 1.19e-13 |
| | | | | R/IDSLLENDR | 537.738 | 1073.535 | 0.037 | 8.74 | 2.17e-13 |
| | | | | R/QQTHALDVM*Q#DSFNR | 903.272 | 1805.8 | 0.636 | 7.75 | 6.29e-10 |
| | | | | R/ASSIM*DELFQDR | 714.215 | 1426.639 | 0.112 | 10.4 | 1.42e-20 |
| | | | | R/EPQDTYHYSPFSLFQR | 1007.856 | 2013.922 | 0.113 | 6.16 | 8.72e-05 |
| 4302 | TSP1_HUMAN | P07996 | 4 | K/GPDPSSPAFR | 515.657 | 1029.488 | 0.095 | 9.93 | 3.01e-18 |
| | | | | R/IEDANLIPPVPDDKFQDLVDAVR | 860.853 | 2578.328 | -0.403 | 8.3 | 1.43e-11 |
| | | | | K/FQDLVDAVR | 531.714 | 1061.55 | 0.069 | 8.82 | 1.02e-13 |
| | | | | K/GGVNDNFQGVLQNVR | 808.804 | 1615.806 | 0.107 | 8.53 | 1.08e-12 |

60

| | | | | | | | | | |
|---|---|---|---|---|---|---|---|---|---|
| 5101 | LMNA_RAT | P48679 | 9 | R/LQEKEDLQELNDR | 815.299 | 1628.8 | 0.109 | 9.73 | 1.57e-17 |
| | | | | R/SLETENAGLR | 545.274 | 1088.546 | 0.007 | 9.19 | 8.26e-15 |
| | | | | R/ITESEEVVSR | 574.713 | 1147.572 | 0.081 | 12.2 | 2.80e-29 |
| | | | | K/AAYEAELGDAR | 582.603 | 1164.541 | 0.675 | 9.6 | 1.52e-16 |
| | | | | R/LKDLEALLNSK | 622.29 | 1242.718 | 0.077 | 9.89 | 3.58e-18 |
| | | | | K/EAALSTALSEKR | 638.275 | 1274.683 | 0.074 | 6.15 | 6.67e-05 |
| | | | | K/EAALSTALSEK | 560.2 | 1118.581 | 0.098 | 9.7 | 2.86e-17 |
| | | | | R/TLEGELHDLR | 591.74 | 1181.604 | 0.07 | 9.18 | 3.50e-15 |
| | | | | R/LQ#TLKEELDFQK | 497.874 | 1491.782 | 0.394 | 7.24 | 5.12e-08 |
| 5106 | CO5A2_HUMAN | P05997 | 3 | K/SLSSQIETM*R | 584.212 | 1166.56 | 0.076 | 10.4 | 2.84e-20 |
| | | | | R/GSQFAYGDHQSPNTAITQM*TFLR | 862.33 | 2585.197 | 0.41 | 6.84 | 1.10e-06 |
| | | | | K/NSVGYM*DDQAK | 622.16 | 1242.518 | 0.107 | 10 | 2.68e-18 |
| 7006 | EF2_RAT | P05197 | 3 | R/ETVSEESNVLC^LSK | 797.785 | 1593.755 | 0.1 | 16.1 | 1.90e-53 |
| | | | | K/ARPFPDGLAEDIDKGEVSAR | 714.301 | 2142.07 | 0.73 | 13.6 | 5.50e-37 |
| | | | | R/C^LYASVLTAQPR | 689.733 | 1377.707 | 0.128 | 11.9 | 2.03e-27 |
| 7510 | VINC_HUMAN | P18206 | 4 | K/ELLPVLISAM*K | 614.446 | 1228.71 | 0.917 | 8.52 | 2.86e-12 |
| | | | | K/Q#VATALQNLQTK | 657.778 | 1314.714 | 0.587 | 8.2 | 1.93e-11 |
| | | | | R/ALASQLQDSLK | 587.247 | 1172.64 | 0.081 | 10.5 | 8.42e-21 |
| | | | | K/AVAGNISDPGLQK | 635.247 | 1268.672 | 0.097 | 9.06 | 1.02e-14 |
| 8110 | SERC_HUMAN | Q9Y617 | 8 | R/QVVNFGPGPAK | 557.281 | 1112.597 | 0.025 | 8.24 | 3.43e-11 |
| | | | | R/ELLAVPDNYK | 580.824 | 1160.607 | 0.487 | 7.51 | 1.14e-08 |
| | | | | K/GAVLVC^DM*SSSNFLSK | 821.989 | 1642.769 | 0.403 | 7.3 | 2.52e-07 |
| | | | | K/GAVLVC^DM*SSNFLSKPVDVSK | 757.019 | 2268.113 | 0.026 | 8.83 | 1.93e-13 |
| | | | | R/DDLLGFALR | 509.705 | 1018.544 | 0.575 | 7.01 | 4.97e-07 |
| | | | | R/EC^PSVLEYK | 562.753 | 1123.522 | 0.016 | 7.92 | 2.03e-10 |
| | | | | K/ALELNM*LSLK | 573.944 | 1146.631 | 0.379 | 9.29 | 2.90e-15 |
| | | | | R/ASLYNAVTIEDVQK | 775.374 | 1549.798 | 0.533 | 9.82 | 6.97e-18 |
| 8111 | ALDOA_HUMAN | P04075 | 5 | R/QLLLTADDR | 522.771 | 1043.561 | 0.017 | 10.3 | 1.20e-19 |
| | | | | R/VNPC^IGGVILFHETLYQK | 696.365 | 2087.087 | 0.338 | 8.26 | 2.44e-11 |
| | | | | K/GVVPLAGTNGETTTQ#GLDGLSER | 758.492 | 2273.102 | 0.216 | 6.71 | 3.00e-06 |
| | | | | K/IGEHTPSALAIM*ENANVLAR | 708.385 | 2122.084 | -0.016 | 8.85 | 1.51e-13 |
| | | | | K/C^PLLKPWALTFSYGR | 905.045 | 1807.944 | -0.065 | 9.57 | 6.92e-17 |

Table 3

**Prediction of dog protein orthologs and functional classification**

All peptide sequence tags (Table 2) were searched against the dog genome database using BLASTP version 2.2.16 (32, 33). Database size was 33,527 dog RefSeq protein sequences. This database is hosted at NCBI. Searches were performed as follows: word size 3, filter low complexity, expect value 0.01, score matrix BLOSUM62. Failed searches were repeated with settings for "short and nearly exact matches" (in bold): word size 2, filter off, expect value 20,000, score matrix PAM30. Only the top scoring significant hit was accepted. Functional classification according to Human Protein Reference Database (34). SSP assigned by image analysis software PDQuest version 7.3.1.

| SSP | Protein description | NCBI accession number | Score | Expect value | Biological process | Molecular function |
|---|---|---|---|---|---|---|
| 502 | Clusterin | NP_001003370 | 99 | 9.00e-22 | Immune response | Complement activity |
| 403/9302 | PREDICTED: similar to Protein-lysine 6-oxidase precursor isoform 3 | XP_859412 | 90.5 | 3.00e-19 | Cell growth and/or maintenance | Catalytic activity |
| 602/9602 | Lectin, mannose-binding 2 | NP_001003258 | 219 | 8.00e-58 | Transport | Transporter activity |
| 1104 | PREDICTED: similar to Collagen alpha 2(V) chain percursor | XP_535998 | 50.4 | 3.00e-07 | Cell growth and/or maintenance | Extracellular matrix structural constituent |
| 1405 | Clusterin | NP_001003370 | **61.3** | **7.00e-11** | Immune response | Complement activity |
| 1502 | Clusterin | NP_001003370 | 125 | 1.00e-29 | Immune response | Complement activity |
| 1602 | Lectin, mannose-binding 2 | NP_001003258 | 129 | 7.00e-31 | Transport | Transporter activity |
| 1703 | PREDICTED: similar to Annexin A1 | XP_533524 | **57.1** | **1.00e-09** | Cell communication; Signal transduction | Calcium ion binding |
| 1802 | PREDICTED: similar to Filamin A isoform 8 | XP_867537 | 60.8 | 2.00e-10 | Cell growth and/or maintenance | Cytoskeletal anchoring activity |
| 2102 | PREDICTED: similar to Nucleophosmin 1 isoform 12 | XP_866781 | 57.8 | 2.00e-09 | Protein metabolism | Chaperone activity |
| 2104 | PREDICTED: similar to Collagen alpha 2(V) chain precursor | XP_535998 | 50.4 | 3.00e-07 | Cell growth and/or maintenance | Extracellular matrix structural constituent |
| 2208 | PREDICTED: similar to Macrophage capping protein | XP_540197 | **48.6** | **4.00e-07** | Cell growth and/or maintenance | Cytoskeletal protein binding |
| 3502 | PREDICTED: | XP_859412 | **33.3** | **0.017** | Cell growth and/or maintenance | Catalytic activity |

| | | | | | |
|---|---|---|---|---|---|
| 4104 | similar to Protein-lysine 6-oxidase precursor isoform 3 PREDICTED: | XP_543238 | 79.7 | 5.00e-16 | Cell communication; Signal transduction | Calcium ion binding |
| 4202 | similar to Stanniocalcin-1 precursor Clusterin | NP_001003370 | 103 | 2.00e-23 | Immune response | Complement activity |
| 4302 | PREDICTED: similar to Thrombospondin 1 precursor | XP_544610 | 72 | 1.00e-13 | Cell growth and/or maintenance | Extracellular matrix structural constituent |
| 5101 | PREDICTED: similar to Lamin A/C isoform 5 | XP_864487 | 104 | 2.00e-23 | Cell growth and/or maintenance | Structural molecule activity |
| 5106 | PREDICTED: similar to Collagen alpha 2(V) chain precursor | XP_535998 | 62.8 | 6.00e-11 | Cell growth and/or maintenance | Extracellular matrix structural constituent |
| 7006 | PREDICTED: similar to Elongation factor 2 | XP_533949 | 62 | 1.00e-10 | Protein metabolism | Translation regulator activity |
| 7510 | PREDICTED: similar to Vinculin | XP_536395 | **42.6** | **7.00e-05** | Cell growth and/or maintenance | Cytoskeletal protein binding |
| 8110 | PREDICTED: similar to Phosphoserine aminotransferase isoform 1 | XP_533520 | 72.4 | 7.00e-14 | Metabolism; Energy pathways | Transaminase activity |
| 8111 | PREDICTED: similar to Fructose-bisphosphate aldolase A isoform 2 | XP_849434 | 117 | 3.00e-27 | Metabolism; Energy pathways | Lyase activity |

### 3.4.4 Detection of neo N- and C-termini in peptide fragments

Phenyx offers the remarkable feature to search for non-tryptic peptides, namely, half-cleaved peptides. Trypsin cleavage specificity is then fixed to the N- or C-terminus. In a degradomic approach this option facilitates the detection of neo N- or C-termini defined by the protease activity. In order to find new peptide ends other than lysine or arginine, peptides must not be identified either C- or N-terminal to the neo peptide. We applied this strategy to all protein database searches performed with Phenyx. Several half-cleaved peptides were detected (Table 2). Half-cleaved peptides may also originate from in-source fragmentation of intact tryptic peptides during the ionization process. Accordingly, the following half-cleaved peptides co-eluted with corresponding intact tryptic peptides following chromatographic separation, TDGNSEHLKR and DGNSEHLKR from protein spots SSP 602/9602 and SSP 1602, respectively, and DNDENEHQLSLR from protein spot SSP 2102. The half-cleaved peptides derived from the sequence stretching over amino acids 137-160 of clusterin (IDSLLENDRQQTHALDVMWDSFNR) found in protein spots SSP 502 and SSP 1502 were chromatographically separated and thus may not refer to in-source fragmentation products. Those half-cleaved products are most probably related to in-gel digestion artefacts because cleavage within this protein sequence stretch by meprin must be excluded due to an overall amino acid sequence coverage of this protein that exceeded amino acid 160. The leguminous lectin-like VIP36 was present in two different protein spots (SSP 1602 and SSP 602/9602) and met our criteria for neo proteins. In both spots the half-cleaved tryptic peptide LFQLMVEH (residues 273-280) was identified and no further peptides towards the C-terminal end. Additionally, this half-cleaved peptide was not generated by in-source fragmentation because there was no co-eluting ion trace of the corresponding fully tryptic peptide LFQLMVEHTPDEENIDWTK. VIP36 was described as a single-pass type I membrane protein with an extracellular carbohydrate recognition domain (CRD) exactly terminating at those amino acids (residues 52-280) (37). Moreover, the amino acid sequence following the putative cleavage site corresponded to cleavage preference for meprin$\alpha$ with the amino acids threonine and proline in the P1' and P2' position (10). The targeted cleavage by hmeprin after this specific domain may point to protein ectodomain shedding. Nevertheless the biological consequence remains to be elucidated.

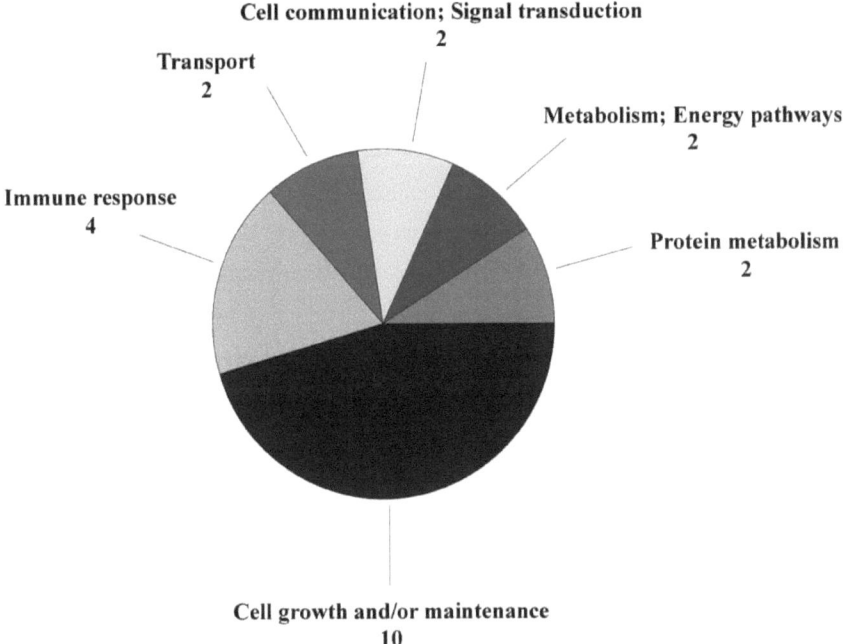

**Fig. 5. Functional classification of identified proteins.**
Pie chart showing distribution of 22 identified proteins into their functional classes. Functional classification was performed according to the Human Protein Reference Database (http://www.hprd.org/) (34). Details are listed in Table 3.

## *3.4.5 Functional clustering into biological process and molecular function*

The ultimate goal of systems biology is to define general rules derived from the outcome of a system-wide study. In the context of degradomics we may postulate to cluster major findings into decisive categories. Therefore the proteins identified by LC-MS/MS as putative meprin substrates were classified into functional groups according to the Human Protein Reference Database (Table 3) (34). Interestingly, ten proteins were related to the biological process of "cell growth and/or maintenance" and four to "immune response" (Fig. 5). The remaining proteins were equally distributed into functional classes such as "transport", "cell communication/signal transduction", "metabolism/energy pathways" and "protein metabolism". In conclusion, these findings imply putative functions for meprin in regulation of cell homeostasis and the extracellular environment, and in immune response.

## 3.4.6 Validation of direct or indirect effects by immunoblotting follow up experiments

Proteomics is a very powerful tool for protease-substrate identification, but the data obtained need to be verified by means of alternative techniques. Western blotting experiments revealed not only direct effects exhibited by meprin's activity status (trypsin treated versus non-treated) but also indirect effects mediated by meprin's overexpression (WT versus meprinα/β). Direct effects were observed for vinculin, lysyl oxidase and collagen type V (Fig. 6A-C). Indirect effects were noticed for annexin A1 (Fig. 6D).

The cytosolic actin-binding protein vinculin was found in the culture medium of MDCK cells (Fig. 6A). The 116 kDa full-length form was detected in all samples whereas putative cleavage products with molecular weights of 75 kDa and 85 kDa (38), respectively, were visualized exclusively in media of MDCK cells expressing active meprinα/β. Vinculin is a representative of the group and subgroup "cell growth and/or maintenance" and "cytoskeletal protein-binding" (Table 3).

The ECM stabilizing protein lysyl oxidase was reported to be synthesized as a 46 kDa precursor that is processed in the extracellular environment to the catalytically functional 32 kDa form by BMP-1/Tolloid (TLD)-like metalloendopeptidases (39). We observed the presence of a 25 kDa protein species in media of MDCK cells expressing active meprinα/β (Fig. 6B).

Type V collagen is a quantitatively minor component of predominantly type I collagen fibrils in most non-cartilage tissues and is required for collagen fibril nucleation (40). The monoclonal antibody 1E2-E4/Col5 did not recognize collagen type V on blots of denaturing, reducing SDS gels (Fig. 6C) (36). Repetition of the experiment under non-denaturing, non-reducing conditions on a dot blot confirmed the absence of native collagen type V in media of wild-type and meprinα/β MDCK cells (data not shown). We then systematically mapped all peptide sequences of collagen type V identified by LC-MS/MS to the full-length sequence as deposited in uniprot-SwissProt protein database. Interestingly, all tryptic peptides from the three independent protein spots SSP 1104, SSP 2104 and SSP 5106 matched the C-terminal propeptide region (residues 1227-1496) of collagen type V (Table 2). This finding corroborates a putative role for hmeprin in the regulation of collagen assembly.

Annexin A1 is a calcium/phospholipid-binding protein providing a link between calcium signalling and membrane functions (41). Two bands 32 kDa and 35 kDa in size were found in conditioned media of wild-type MDCK cells (Fig. 6D). In media of MDCKα/β cells the 32 kDa form was not detectable. Obviously, overexpression of hmeprinα/β in MDCK cells abolished the 32 kDa band. There was no marked difference between the trypsin treated and non-treated cells. This finding may indicate an indirect effect exerted by meprin's overexpression per se and not by meprin's activity status.

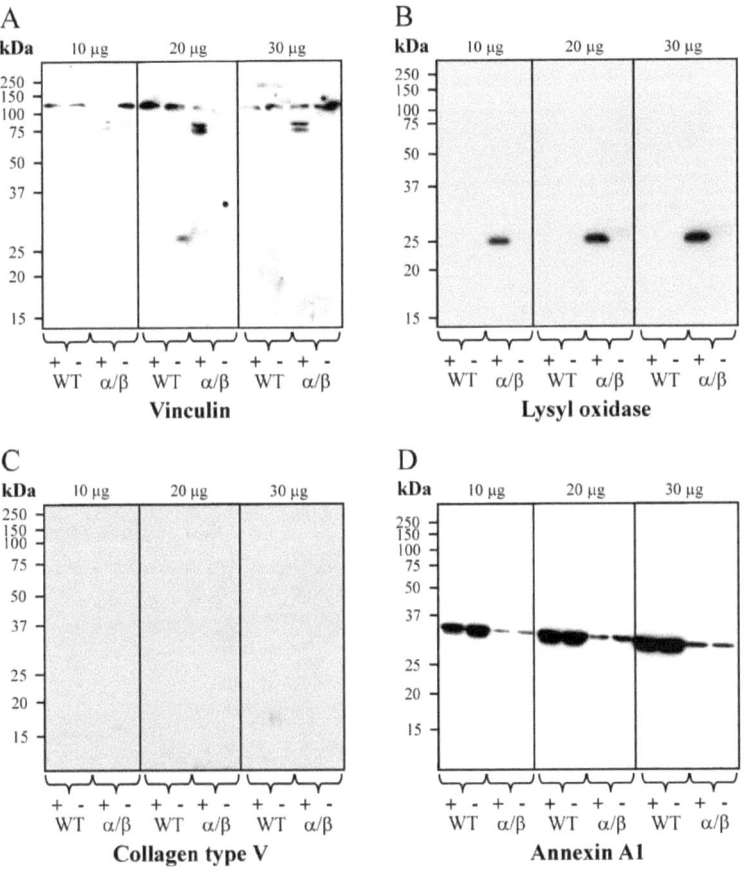

**Fig. 6. Validation experiments by western blotting.**
Conditioned medium protein of trypsin treated (+) and non-treated (-) wild-type and meprinα/β MDCK cells was separated according to mass as described in Figure 1. Immunoblotting with antibodies against (A) vinculin, (B) lysyl oxidase, (C) collagen type V and (D) annexin A1. Migration positions of molecular mass standards and protein loading amounts are indicated.

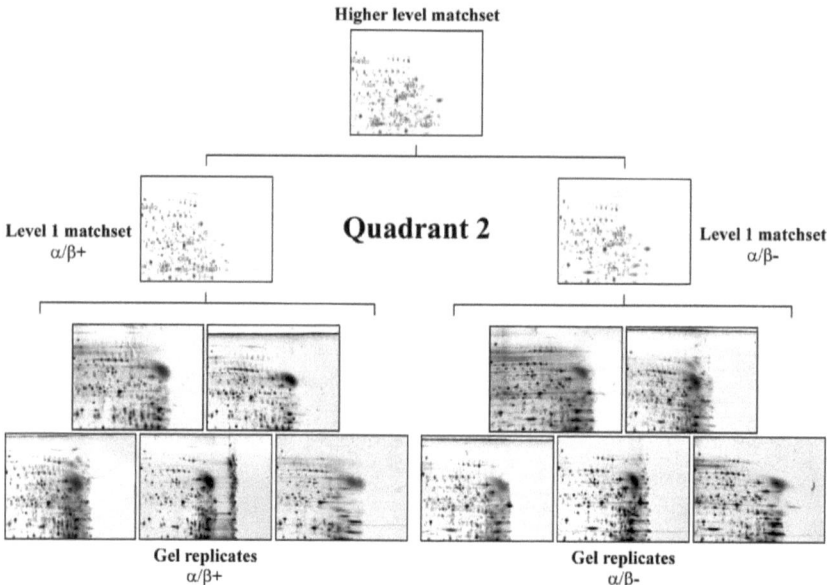

**Fig. S1. Two-dimensional gel image analysis of the second quadrant section.**

250 µg of conditioned medium protein from trypsin treated (+) and non-treated (-) MDCKα/β cells was separated by IEF in a 24 cm long IPG pH 3-10 NL strip. Vertical separation according to mass in a 12.5% SDS gel. Optimized Ruthenium staining. For each condition three pooled biological gel replicates (from 18 dishes per pooled sample) and two more technical gel replicates (of one pooled sample) were produced for subsequent image analysis. Unique protein spots are labelled in level 1 and higher level matchsets with SSP assigned by the image analysis software.

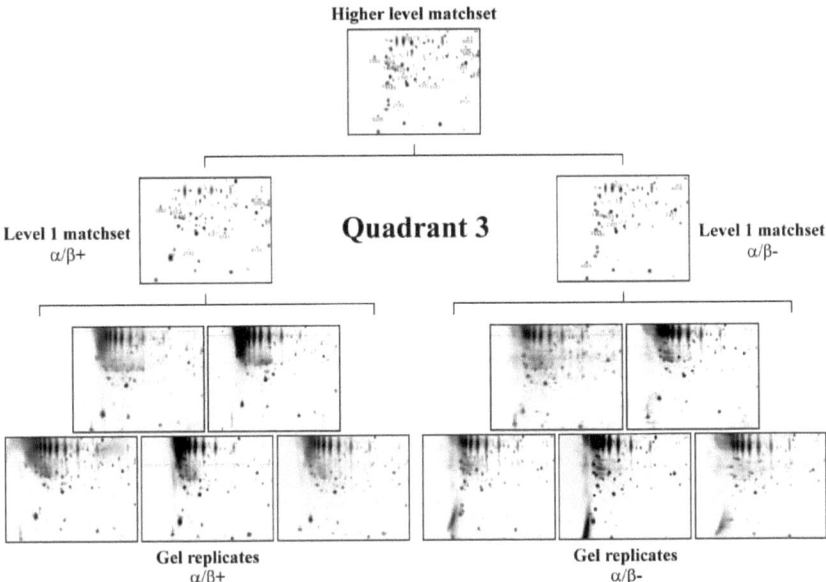

**Fig. S2. Two-dimensional gel image analysis of the third quadrant section.**

250 μg of conditioned medium protein from trypsin treated (+) and non-treated (-) MDCKα/β cells was separated by IEF in a 24 cm long IPG pH 3-10 NL strip. Vertical separation according to mass in a 12.5% SDS gel. Optimized Ruthenium staining. For each condition three pooled biological gel replicates (from 18 dishes per pooled sample) and two more technical gel replicates (of one pooled sample) were produced for subsequent image analysis. Unique protein spots are labelled in level 1 and higher level matchsets with SSP assigned by the image analysis software.

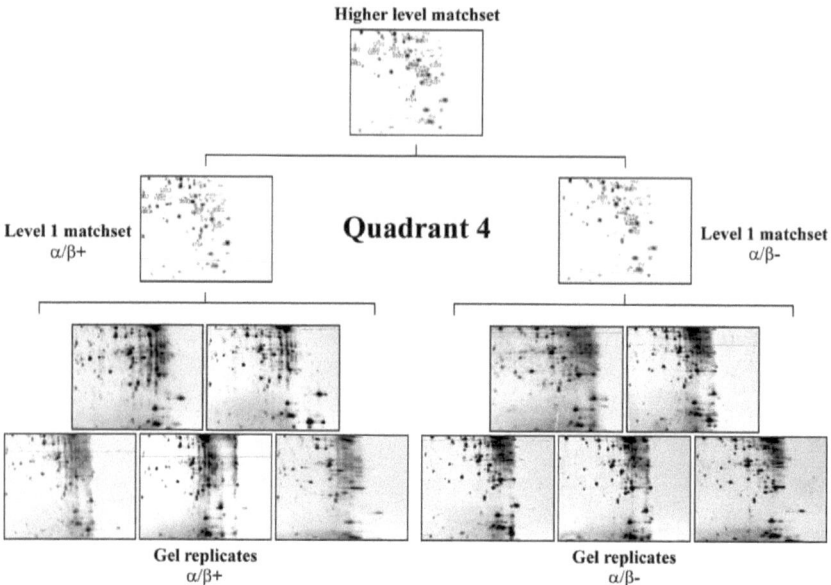

**Fig. S2. Two-dimensional gel image analysis of the fourth quadrant section.**

250 µg of conditioned medium protein from trypsin treated (+) and non-treated (-) MDCKα/β cells was separated by IEF in a 24 cm long IPG pH 3-10 NL strip. Vertical separation according to mass in a 12.5% SDS gel. Optimized Ruthenium staining. For each condition three pooled biological gel replicates (from 18 dishes per pooled sample) and two more technical gel replicates (of one pooled sample) were produced for subsequent image analysis. Unique protein spots are labelled in level 1 and higher level matchsets with SSP assigned by the image analysis software.

## 3.5 Discussion

### 3.5.1 Establishment of a novel two-dimensional gel-based degradomic approach

To date, MMPs and ADAMs, two metzincin metalloendopeptidase families have been characterized on a system-wide level by means of degradomics (16-19). Two degradomic approaches defined the substrate repertoire of membrane-type 1-MMP (MT1-MMP/MMP-14) in a cell culture system-based environment or using human plasma as a polysubstrate (17, 19). The other two reports described the substrate degradome of TACE/ADAM-17 in a cell culture system (16, 18). These degradomes were defined using multi-dimensional LC-MS/MS with ICAT labelling or 2-DE with (or without) lectin-affinity pre-fractionation and cyanine dye labelling. Interestingly, none of these studies systematically grouped the putative substrates into specific, functional categories to draw meaningful conclusions on a whole system-wide level. In addition, no systematic display of data on biological replicates was represented. Moreover, applying pre-fractionation techniques, namely, selection for cysteine-containing peptides or glycoproteins, may allow for higher resolution capacity but at the cost of information loss. For example, ICAT labelling does not allow detecting neo N- or C-termini of cleavage products and non-glycosylated proteins and fragments escape from lectin-affinity purification.

Here, we clearly demonstrate the applicability of a novel 2-D gel image analysis strategy to define the hmeprin substrate degradome in a cell culture system-based approach (Fig. 2, Fig. 3, supplementary Fig. S1-S3 and Table 1). Despite previous reports on limited resolution capacity of 2-D gels to find putative cleavage candidates other than heat shock proteins, actin and metabolic pathway enzymes, we were able to identify novel substrates without the use of sophisticated pre-fractionation techniques (16, 18). The main success of our findings may be attributed to improved detection sensitivity of Ruthenium staining as the initial 2-D gel-based investigation applied a commercially available colloidal Coomassie stain lacking sensitivity of our house-made fluorescent dye (16). Additionally, the sample preparation protocol described herein was compatible with DIGE technology and may direct future applications. Another major progress was achieved by restricting the large-scale proteomic study to media of trypsin treated versus non-treated MDCK$\alpha/\beta$ cells and by extensive replication on the 2-D gel level (three pooled biological gel replicates and two more technical gel replicates). The systematic production of 2-D gel sets enabled the design of a novel image analysis strategy that unmasked step-by-step qualitative differences on each level of perspective. The modular character of those building blocks allowed the integration of the whole system's information into one big differential reference map. Due to the poor representation of dog proteins (664 sequence entries) in the uniprot-SwissProt protein database (release 51.3) unambiguous protein and species identifications were inferred from peptide sequence tags searched with BLASTP version 2.2.16 against NCBI's dog genome database (33,527 dog RefSeq protein sequence entries) (Table 3) (32, 33). This valuable strategy was used in similar cases where the species of interest was underrepresented in protein databases (42). Moreover, visual inspection of the peptide sequence list revealed a hitherto unkown half-cleaved tryptic peptide, namely, LFQLMVEH (residues 273-280) of VIP36 (Table 2), that perfectly matched cleavage preference for meprinα (10). This peptide finding pointed to a putative neo C-terminus and may overcome technical limitations of the ICAT-based approach (17). Finally, classification of protein identifications into decisive functional groups with HPRD facilitated the system-wide interpretation of our data (Table 3 and Fig. 5) (34). Hmeprin's pivotal functions were then assigned to "cell growth and/or maintenance" and to "immune response". Altogether, the novel strategies and applications presented herein may help to understand more precisely the function of a protease in a complex environment.

## 3.5.2 Key roles for human meprin in homeostasis of cell, cellular environment and in immune response

BMP-1, mammalian tolloid (mTLD) and hmeprin belong to the same metzincin subfamily of metalloendopeptidases, the astacin family (43). The main functions of BMP-1/TLD-like metalloendopeptidases are ascribed to the proteolytic removal of C-propeptides from fibrillar procollagens and to activation of lysyl oxidase (39, 40, 44). Upon activation, lysyl oxidase mediates the oxidative deamination of lysine residues to highly reactive aldehydes that spontaneously cross-link processed collagen monomers (39). Cross-linkage in self-assembling fibrous collagen is essential for its structural integrity. Quantitatively minor type V collagen mainly exists as $\alpha1(V)_2\alpha2(V)$ heterotrimer that is incorporated into type I collagen fibrils and initiates collagen fibril assembly in regions of new tissue formation (36, 40, 44). Interestingly, we identified collagen type V in three individual protein spots, probably differently glycosylated protein isoforms, and all peptides identified exclusively matched the C-terminal propeptide region (residues 1227-1496) (Table 2). In addition, the monoclonal antibody 1E2-E4/Col5 did not detect native collagen type V in cell culture supernatants of wild-type and meprinα/β MDCK cells under non-denaturing, non-reducing conditions. Moreover, a single 25 kDa protein form of lysyl oxidase was exclusively found in media of trypsin treated MDCKα/β cells (Fig. 6B). As previously described, lysyl oxidase acts solely on processed collagens and not on its precursors, thus the 25 kDa form most likely exhibits amine oxidase activity (39). Hence we speculate that hmeprin has activity similar to BMP-1/TLD-like metalloendopeptidases in that it acts as a procollagen C protease as well as an activator of lysyl oxidase. An important role for hmeprinα/β may therefore be in tissue remodelling processes through the targeted regulation of ECM assembly and not by aberrant destruction.

Vinculin is an actin-binding protein localized on the cytoplasmic face of integrin-mediated cell-ECM junctions designated as focal adhesions (38). Vinculin stabilizes focal adhesions and thereby suppresses cell migration. This effect is relieved by transient changes in local concentrations of inositol phospholipids. It thus serves a regulatory, dynamic linkage between the ECM and intracellular actin cytoskeleton. It was demonstrated that acidic phospholipids inhibit intramolecular association between the N- and C-terminal regions of vinculin, exposing actin-binding and PKC phosphorylation sites (45), namely, serines 1033 and 1045 (46). Upon activation of hmeprinα/β in stably transfected MDCK cells the 116 kDa full-length form of vinculin and truncated forms (75 kDa and 85 kDa) were detected in cell culture supernatants (Fig. 6A). These findings raise the question how meprin elicits such effects as the catalytic protease domain is localized extracellularly (43). A reasonable explanation is that meprin may exert its functions by intracellular signalling. Indeed hmeprinβ possesses a C-terminal cytoplasmic domain with a PKC consensus sequence that may be phosphorylated on serine 687 upon PMA (47). Hence meprinβ, the only astacin family member that is membrane-bound, may also function as a signalling receptor (43). This unique feature may provide a link to another metzincin subfamily, the ADAMs, since ADAM-15 was reported to interact specifically with Src family protein-tyrosine kinases upon phosphorylation on tyrosines 715 and 735 (48). Although the exact mechanisms underlying secondary or downstream intracellular proteolytic events are not clear, hmeprin may also mediate important intracellular signalling via its C-terminal domain ultimately leading to regulation of cytoskeletal rearrangement during tissue remodelling processes.

The leguminous lectin-like vesicular integral-membrane protein VIP36 was originally identified as a component of glycolipid rafts and exocytic carrier vesicles in epithelial cells (49). Due to its homology to the mannose-selective lectin ERGIC-53 it was suggested that VIP36 may operate in quality control of glycosylation in the Golgi (50). However, in MDCK cells VIP36 is also localized to the apical plasma membrane and appears to be involved in intracellular transport and secretion of glycoproteins containing N-linked glycans (51). The meprins are extensively glycosylated consisting of ~25% carbohydrates which are N-linked in meprinα and both N- and O-linked in meprinβ (52, 53). Hence VIP36 could potentially interact with meprinα and/or meprinβ via N-linked glycans and this may facilitate apical protein sorting to the plasma membrane. As there is no one specific glycosylation site or type of oligosaccharide (high mannose- or complex-type) that determines apical sorting of mouse meprinα (52), VIP36 may direct apical targeting. Upon detailed analysis of peptides from two separate protein spots corresponding to VIP36 we could consistently match all peptide sequences to the extracellular CRD (Table 2). Since VIP36 is a single-pass type I membrane protein, and half-cleaved peptides terminating exactly at the end of the CRD were found, hmeprinα may shed VIP36 from the plasma membrane (37). Analogous protein ectodomain shedding processes were described for MT-MMPs and ADAMs (16-18). The biological consequence of this event remains elusive.

In conclusion, since the introduction of the term degradome various novel technological approaches emerged that were successfully applied to decipher the substrate repertoire of a given protease on a system-wide level (15-19). The work presented herein is the first degradomic approach implemented on an astacin family member of the metzincin superfamily of metalloendopeptidases. Upon our findings hmeprin may be considered as a signalling protease mediating direct and indirect cleavage and signal transduction functions rather than aberrantly destroying the ECM. Although the detailed mechanisms need to be extensively studied, hmeprin is preferentially involved in "cell growth and/or maintenance" and in "immune response". Fascinatingly, all degradomic approaches employed on metzincin metalloendopeptidases, namely, MMP-14 and TACE/ADAM-17 (16-19), allowed the authors to come to the same conclusion: metalloendopeptidases are most crucial entities in regulation of cell homeostasis and its environment and in innate immunity. Therefore future systems biology studies may expand our current knowledge on astacin family members and its relatives.

## 3.6 References

1. Sterchi, E. E., Green, J. R., Lentze, M. J. (1982) Non-pancreatic hydrolysis of N-benzoyl-l-tyrosyl-p-aminobenzoic acid (PABA-peptide) in the human small intestine. *Clin. Sci.* 62, 557-560
2. Sterchi, E. E., Naim, H. Y., Lentze, M. J., Hauri, H. P., Fransen, J. A. (1988) N-benzoyl-L-tyrosyl-p-aminobenzoic acid hydrolase: a metalloendopeptidase of the human intestinal microvillus membrane which degrades biologically active peptides. *Arch. Biochem. Biophys.* 265, 105-118
3. Beynon, R. J, Shannon, J. D, Bond, J. S. (1981) Purification and characterization of a metallo-endoproteinase from mouse kidney. *Biochem. J.* 199, 591-598
4. Kenny, A. J., Fulcher, I. S., Ridgwell, K., Ingram, J. (1981) Microvillar membrane neutral endopeptidases. *Acta Biol. Med. Ger.* 40, 1465-1471
5. Grünberg, J., Dumermuth, E., Eldering, J. A., Sterchi, E. E. (1993) Expression of the alpha subunit of PABA peptide hydrolase (EC 3.4.24.18) in MDCK cells. Synthesis and secretion of an enzymatically inactive homodimer. *FEBS Lett.* 335, 376-379
6. Eldering, J. A., Grünberg, J., Hahn, D., Croes, H. J., Fransen, J. A., Sterchi, E. E. (1997) Polarised expression of human intestinal N-benzoyl-L-tyrosyl-p-aminobenzoic acid hydrolase (human meprin) alpha and beta subunits in Madin-Darby canine kidney cells. *Eur. J. Biochem.* 247, 920-932
7. Rösmann, S., Hahn, D., Lottaz, D., Kruse, M. N., Stöcker, W., Sterchi, E. E. (2002) Activation of human meprin-alpha in a cell culture model of colorectal cancer is triggered by the plasminogen-activating system. *J. Biol. Chem.* 277, 40650-40658
8. Becker, C., Kruse, M. N., Slotty, K. A., Köhler, D., Harris, J. R., Rösmann, S., Sterchi, E. E., Stöcker, W. (2003) Differences in the activation mechanism between the alpha and beta subunits of human meprin. *Biol. Chem.* 384, 825-831
9. Yamaguchi, T., Fukase, M., Kido, H., Sugimoto, T., Katunuma, N., Chihara, K. (1994) Meprin is predominantly involved in parathyroid hormone degradation by the microvillar membranes of rat kidney. *Life Sci.* 54, 381-386
10. Bertenshaw, G. P., Turk, B. E., Hubbard, S. J., Matters, G. L., Bylander, J. E., Crisman, J. M., Cantley, L. C., Bond, J. S. (2001) Marked differences between metalloproteases meprin A and B in substrate and peptide bond specificity. *J. Biol. Chem.* 276, 13248-13255
11. Kaushal, G. P., Walker, P. D., Shah, S. V. (1994) An old enzyme with a new function: purification and characterization of a distinct matrix-degrading metalloproteinase in rat kidney cortex and its identification as meprin. *J. Cell Biol.* 126, 1319-1327
12. Walker, P. D., Kaushal, G. P., Shah, S. V. (1998) Meprin A, the major matrix degrading enzyme in renal tubules, produces a novel nidogen fragment in vitro and in vivo. *Kidney Int.* 53, 1673-1680
13. Herzog, C., Kaushal, G. P., Haun, R. S. (2005) Generation of biologically active interleukin-1beta by meprin B. *Cytokine* 31, 394-403
14. Villa, J. P., Bertenshaw, G. P., Bylander, J. E., Bond, J. S. (2003) Meprin proteolytic complexes at the cell surface and in extracellular spaces. *Biochem. Soc. Symp.* 53-63
15. Lopez-Otin, C., Overall, C. M. (2002) Protease degradomics: a new challenge for proteomics. *Nat. Rev. Mol. Cell Biol.* 3, 509-519

16. Guo, L., Eisenman, J. R., Mahimkar, R. M., Peschon, J. J. Paxton, R. J., Black, R. A., Johnson, R. S. (2002) A proteomic approach for the identification of cell-surface proteins shed by metalloproteases. *Mol. Cell. Proteomics* 1, 30-36

17. Tam, E. M., Morrison, C. J., Wu, Y. I., Stack, M. S., Overall, C. M. (2004) Membrane protease proteomics: Isotope-coded affinity tag MS identification of undescribed MT1-matrix metalloproteinase substrates. *Proc. Natl. Acad. Sci. U S A* 101, 6917-6922

18. Bech-Serra, J. J., Santiago-Josefat, B., Esselens, C., Saftig, P., Baselga, J., Arribas, J., Canals, F. (2006) Proteomic identification of desmoglein 2 and activated leukocyte cell adhesion molecule as substrates of ADAM17 and ADAM10 by difference gel electrophoresis. *Mol. Cell. Biol.* 26, 5086-5095

19. Hwang, I. K., Park, S. M., Kim, S. Y., Lee, S. T. (2004) A proteomic approach to identify substrates of matrix metalloproteinase-14 in human plasma. *Biochim. Biophys. Acta* 1702, 79-87

20. Overall, C. M., Dean, R. A. (2006) Degradomics: systems biology of the protease web. Pleiotropic roles of MMPs in cancer. *Cancer Metastasis Rev.* 25, 69-75

21. Overall, C. M. (2004) Dilating the degradome: Matrix metalloproteinase 2 (MMP-2) cuts to the heart of the matter. *Biochem. J.* 383, e5-e7

22. Parks, W. C., Wilson, C. L., Lopez-Boado, Y. S. (2004) Matrix metalloproteinases as modulators of inflammation and innate immunity. *Nat. Rev. Immunol.* 4, 617-629

23. Richardson, J. C., Scalera, V., Simmons, N. L. (1981) Identification of two strains of MDCK cells which resemble separate nephron tubule segments. *Biochim. Biophys. Acta* 673, 26-36

24. Ambort, D., Lottaz, D., Sterchi, E. (2007) Sample preparation of culture medium from Madin-Darby canine kidney cells. In *Methods in Molecular Biology: Sample Preparation and Pre-Fractionation for 2-D PAGE: Methods and Protocols in Expression Proteomics (Posch, A., ed.)*. Humana Press, Totowa, NJ (in press)

25. Laemmli, U. K. (1970) Cleavage of structural proteins during the assembly of the head of bacteriophage T4. *Nature* 15, 680-685

26. Hoving, S., Voshol, H., van Oostrum, J. (2005) Using ultra-zoom gels for high resolution two-dimensional polyacrylamide gel electrophoresis. In *Methods in Molecular Biology: The Proteomics Protocols Handbook (Walker, J.M., ed.)*. Humana Press, Totowa, NJ, pp. 151-166

27. Rabilloud, T., Strub, J. M., Luche, S., van Dorsselaer, A., Lunardi, J. (2001) A comparison between Sypro Ruby and ruthenium II tris (bathophenanthroline disulfonate) as fluorescent stains for protein detection in gels. *Proteomics* 1, 699-704

28. Lamanda, A., Zahn, A., Röder, D., Langen, H. (2004) Improved Ruthenium II tris (bathophenanthroline disulfonate) staining and destaining protocol for a better signal-to-background ratio and improved baseline resolution. *Proteomics* 4, 599-608

29. Anderson, N. L., Esquer-Blasco, R., Hofmann, J. P., Anderson, N. G. (1991) A two-dimensional gel database of rat liver proteins useful in gene regulation and drug effects studies. *Electrophoresis* 12, 907-930

30. Heller, M., Stalder, D., Schlappritzi, E., Hayn, G., Matter, U., Haeberli, A. (2005) Mass spectrometry-based analytical tools for the molecular protein characterization of human plasma lipoproteins. *Proteomics* 5, 2619-2630

31. Perkins, D. N., Pappin, D. J., Creasy, D. M., and Cottrell, J. S. (1999) Probability-based protein identification by searching sequence databases using mass spectrometry data. *Electrophoresis* 20, 3551-3567

32. Altschul, S. F., Madden, T. L., Schaffer, A. A., Zhang, J., Zhang, Z., Miller, W., Lipman, D. J. (1997) Gapped BLAST and PSI-BLAST: a new generation of protein database search programs. *Nucleic Acids Res.* 25, 3389-3402

33. Lindblad-Toh, K., Wade, C. M., Mikkelsen, T.S., Karlsson, E. K., Jaffe, D. B., Kamal, M., Clamp, M., Chang, J. L., Kulbokas, E. J. 3rd., Zody, M. C., Mauceli, E., Xie, X., Breen, M., Wayne, R. K., Ostrander, E. A., Ponting, C. P., Galibert, F., Smith, D. R., DeJong, P. J., Kirkness, E., Alvarez, P., Biagi, T., Brockman, W., Butler, J., Chin, C. W., Cook, A., Cuff, J., Daly, M.J., DeCaprio, D., Gnerre, S., Grabherr, M., Kellis, M., Kleber, M., Bardeleben, C., Goodstadt, L., Heger, A., Hitte, C., Kim, L., Koepfli, K. P., Parker, H. G., Pollinger, J. P., Searle, S. M., Sutter, N. B., Thomas, R., Webber, C., Baldwin, J., Abebe, A., Abouelleil, A., Aftuck, L., Ait-Zahra, M., Aldredge, T., Allen, N., An, P., Anderson, S., Antoine, C., Arachchi, H., Aslam, A., Ayotte, L., Bachantsang, P., Barry, A., Bayul, T., Benamara, M., Berlin, A., Bessette, D., Blitshteyn, B., Bloom, T., Blye, J., Boguslavskiy, L., Bonnet, C., Boukhgalter, B., Brown, A., Cahill, P., Calixte, N., Camarata, J., Cheshatsang, Y., Chu, J., Citroen, M., Collymore, A., Cooke, P., Dawoe, T., Daza, R., Decktor, K., DeGray, S., Dhargay, N., Dooley, K., Dooley, K., Dorje, P., Dorjee, K., Dorris, L., Duffey, N., Dupes, A., Egbiremolen, O., Elong, R., Falk, J., Farina, A., Faro, S., Ferguson, D., Ferreira, P., Fisher, S., FitzGerald, M., Foley, K., Foley, C., Franke, A., Friedrich, D., Gage, D., Garber, M., Gearin, G., Giannoukos, G., Goode, T., Goyette, A., Graham, J., Grandbois, E., Gyaltsen, K., Hafez, N., Hagopian, D., Hagos, B., Hall, J., Healy, C., Hegarty, R., Honan, T., Horn, A., Houde, N., Hughes, L., Hunnicutt, L., Husby, M., Jester, B., Jones, C., Kamat, A., Kanga, B., Kells, C., Khazanovich, D., Kieu, A. C., Kisner, P., Kumar, M., Lance, K., Landers, T., Lara, M., Lee, W., Leger, J. P., Lennon, N., Leuper, L., LeVine, S., Liu, J., Liu, X., Lokyitsang, Y., Lokyitsang, T., Lui, A., Macdonald, J., Major, J., Marabella, R., Maru, K., Matthews, C., McDonough, S., Mehta, T., Meldrim, J., Melnikov, A., Meneus, L., Mihalev, A., Mihova, T., Miller, K., Mittelman, R., Mlenga, V., Mulrain, L., Munson, G., Navidi, A., Naylor, J., Nguyen, T., Nguyen, N., Nguyen, C., Nguyen, T., Nicol, R., Norbu, N., Norbu, C., Novod, N., Nyima, T., Olandt, P., O'Neill, B., O'Neill, K., Osman, S., Oyono, L., Patti, C., Perrin, D., Phunkhang, P., Pierre, F., Priest, M., Rachupka, A., Raghuraman, S., Rameau, R., Ray, V., Raymond, C., Rege, F., Rise, C., Rogers, J., Rogov, P., Sahalie, J., Settipalli, S., Sharpe, T., Shea, T., Sheehan, M., Sherpa, N., Shi, J., Shih, D., Sloan, J., Smith, C., Sparrow, T., Stalker, J., Stange-Thomann, N., Stavropoulos, S., Stone, C., Stone, S., Sykes, S., Tchuinga, P., Tenzing, P., Tesfaye, S., Thoulutsang, D., Thoulutsang, Y., Topham, K., Topping, I., Tsamla, T., Vassiliev, H., Venkataraman, V., Vo, A., Wangchuk, T., Wangdi, T., Weiand, M., Wilkinson, J., Wilson, A., Yadav, S., Yang, S., Yang, X., Young, G., Yu, Q., Zainoun, J., Zembek, L., Zimmer, A., Lander, E. S. (2005) Genome sequence, comparative analysis and haplotype structure of the domestic dog. *Nature* 438, 803-819

34. Peri, S., Navarro, J. D., Amanchy, R., Kristiansen, T. Z., Jonnalagadda, C. K., Surendranath, V., Niranjan, V., Muthusamy, B., Gandhi, T. K., Gronborg, M., Ibarrola, N., Deshpande, N., Shanker, K., Shivashankar, H. N., Rashmi, B. P., Ramya, M. A., Zhao, Z., Chandrika, K. N., Padma, N., Harsha, H. C., Yatish, A. J., Kavitha, M. P., Menezes, M., Choudhury, D. R., Suresh, S., Ghosh, N., Saravana, R., Chandran, S., Krishna, S., Joy, M., Anand, S. K., Madavan, V., Joseph, A., Wong, G. W., Schiemann, W.

P., Constantinescu, S. N., Huang, L., Khosravi-Far, R., Steen, H., Tewari, M., Ghaffari, S., Blobe, G. C., Dang, C. V., Garcia, J. G., Pevsner, J., Jensen, O. N., Roepstorff, P., Deshpande, K. S., Chinnaiyan, A. M., Hamosh, A., Chakravarti, A., Pandey, A. (2003) Development of human protein reference database as an initial platform for approaching systems biology in humans. *Genome Res.* 13, 2363-2371

35. Towbin, H., Staehelin, T., Gordon, J. (1979) Electrophoretic transfer of proteins from polyacrylamide gels to nitrocellulose sheets: procedure and some applications. *Proc. Natl. Acad. Sci. USA* 76, 4350-4354

36. Werkmeister, J. A., Ramshaw, J. A. (1991) Monoclonal antibodies to type V collagen for immunohistological examination of new tissue deposition associated with biomaterial implants. *J. Histochem. Cytochem.* 39, 1215-1220

37. Neve, E. P., Svensson, K., Fuxe, J., Pettersson, R. F. (2003) VIPL, a VIP36-like membrane protein with a putative function in the export of glycoproteins from the endoplasmic reticulum. *Exp. Cell Res.* 288, 70-83

38. Ziegler, W. H., Liddington, R. C., Critchley, D. R. (2006) The structure and regulation of vinculin. *Trends Cell Biol.* 16, 453-460

39. Lucero, H. A., Kagan, H. M. (2006) Lysyl oxidase: an oxidative enzyme and effector of cell function. *Cell. Mol. Life Sci.* 63, 2304-2316

40. Wenstrup, R. J., Florer, J. B., Brunskill, E. W., Bell, S. M., Chervoneva, I., Birk, D. E. (2004) Type V collagen controls the initiation of collagen fibril assembly. *J. Biol. Chem.* 279, 53331-53337

41. Gerke, V., Creutz, C. E., Moss, S. E. (2005) Annexins: linking Ca2+ signalling to membrane dynamics. *Nat. Rev. Mol. Cell Biol.* 6, 449-461

42. Sunyaev, S., Liska, A. J., Golod, A., Shevchenko, A., Shevchenko, A. (2003) MultiTag: multiple error-tolerant sequence tag search for the sequence-similarity identification of proteins by mass spectrometry. *Anal. Chem.* 15, 1307-15

43. Gomis-Rüth, F. X. (2003) Structural aspect of the metzincin clan of metalloendopeptidases. *Mol. Biotechnol.* 24, 157-202

44. Unsöld, C., Pappano, W. N., Imamura, Y., Steiglitz, B. M., Greenspan, D. S. (2002) Biosynthetic processing of the pro-alpha 1(V)2pro-alpha 2(V) collagen heterotrimer by bone morphogenetic protein-1 and furin-like proprotein convertases. *J. Biol. Chem.* 277, 5596-5602

45. Weekes, J., Barry, S. T., Critchley, D. R. (1996) Acidic phospholipids inhibit the intramolecular association between the N- and C-terminal regions of vinculin, exposing actin-binding and protein kinase C phosphorylation sites. *Biochem. J.* 314, 827-832

46. Ziegler, W. H., Tigges, U., Zieseniss, A., Jockusch, B. M. (2002) A lipid-regulating docking site on vinculin for protein kinase C. *J. Biol. Chem.* 277, 7396-7404

47. Hahn, D., Pischitzis, A., Rösmann, S., Hansen, M. K., Leuenberger, B., Luginbühl, U., Sterchi, E. E. (2003) Phorbol 12-myristate 13-acetate-induced ectodomain shedding and phosphorylation of the human meprinbeta metalloprotease. *J. Biol. Chem.* 278, 42829-42839

48. Poghosyan, Z., Robbins, S. M., Houslay, M. D., Webster, A., Murphy, G., Edwards, D. R. (2002) Phosphorylation-dependent interactions between ADAM15 cytoplasmic domain and Src family protein-tyrosine kinases. *J. Biol. Chem.* 277, 4999-5007

49. Fiedler, K., Parton, R. G., Kellner, R., Etzold, T., Simons, K. (1994) VIP36, a novel component of glycolipid rafts and exocytic carrier vesicles in epithelial cells. *EMBO J.* 13, 1729-1740

50. Hauri, H., Appenzeller, C., Kuhn, F., Nufer, O. (2000) Lectins and traffic in the secretory pathway. *FEBS Lett.* 476, 32-37
51. Hara-Kuge, S., Ohkura, T., Ideo, H., Shimada, O., Atsumi, S., Yamashita, K. (2002) Involvement of VIP36 in intracellular transport and secretion of glycoproteins in polarized Madin-Darby canine kidney (MDCK) cells. *J. Biol. Chem.* 277, 16332-16339
52. Kadowaki, T., Tsukuba, T., Bertenshaw, G. P., Bond, J. S. (2000) N-linked oligosaccharides on the meprin A metalloprotease are important for secretion and enzymatic activity, but not for apical targeting. *J. Biol. Chem.* 275, 25577-25584
53. Leuenberger, B., Hahn, D., Pischitzis, A., Hansen, M. K., Sterchi, E. E. (2003) Human meprin beta: O-linked glycans in the intervening region of the type I membrane protein protect the C-terminal region from proteolytic cleavage and diminish its secretion. *Biochem. J.* 369, 659-665

# Chapter 4

## 4. The substrate degradome of meprin beta in human milk[3]

### 4.1 Summary

In recent years the introduction and application of degradomic techniques led to the discovery that metalloendopeptidases are molecular switches in signalling circuits and not extracellular matrix destroying bulldozers. Here, we report a degradomic approach to determine the substrate repertoire of the astacin-like metalloendopeptidase meprin beta in human milk by combining *in vitro* cleavage assays, one-dimensional gel electrophoresis and liquid chromatography-based tandem mass spectrometry for subsequent protein identification. Of 16 putative candidates prone to hydrolysis by meprin beta, 12 proteins were successfully grouped into decisive functional categories; namely, six into "transport", two into "metabolism/energy pathways" and four into "cell communication/signal transduction". The availability of a complete set of domain-specific monoclonal antibodies against tenascin-C, a representative of the latter group, allowed for accurate mapping of cleavage sites targeted by meprin. For simplicity, a chick tenascin-C model system featuring three differentially spliced variants was used to dissect the substrate specificities exhibited by the two meprin subunits α and β. Interestingly, both meprin subunits processed all three tenascin-C forms by removal of N-terminal 10 kDa and C-terminal 40 kDa peptides but with distinct cleavage preferences. Meprinα preferably attacked the N-terminus whereas meprinβ processed the N- and C-terminus at same efficiency. These findings point to diverse functions of the two meprin subunits: a cleavage in the N-terminal region may disrupt tenascin-C oligomers into monomers whereas C-terminal removal of 40 kDa peptides located within the cell-binding site (seventh fibronectin type III repeat) of tenascin-C may cause detachment from cells and other putative binding partners.

---

[3] Part of this book chapter is prepared for publication in: "Ambort D, Brellier F, Becker-Pauly C, Stöcker W, Chiquet M and Sterchi EE (2009) Specific processing of tenascin-C by the metalloprotease meprinβ neutralizes its inhibition of cell spreading. *Manuscript*."

## 4.2 Introduction

Hmeprin was identified as an astacin-like metalloendopeptidase of human small intestinal mucosa capable of hydrolyzing N-benzoyl-L-tyrosyl-p-aminobenzoic acid, a substrate used for clinical purposes to assess exocrine function of the pancreas (1, 2). At the same time two other groups reported on discovery of meprin (metal endopeptidase from renal tissue) and endopeptidase-2 in mouse and rat kidney (3, 4). Subsequent studies revealed the subunit composition and oligomeric state of meprins. Hmeprin features two forms, named α and β, that are secreted or membrane-bound and may build up disulfide-bonded homo- or heterodimeric complexes; further oligomerization is mediated by non-covalent interactions (5, 6). Both zymogenic forms may be activated by trypsin or alternatively by plasmin for hmeprinα and by kallikrein-4 for hmeprinβ (5, 7-9). With the heterologous overexpression of recombinant hmeprinα and β (rhmeprinα and β) in baculovirus-infected insect cells valuable enzymatic tools were available to elucidate meprin's putative functions by *in vitro* cleavage assays (8, 10).

Traditionally, potential substrates were tested by digestion of selective targets with a protease of interest. A plethora of peptide and protein substrates were subjected to proteolysis by meprin *in vitro*; namely, biologically active peptides (bradykinin, angiotensins I and II (2)), polypeptide hormones (insulin B-chain, parathyroid hormone (11)), ECM components (collagen type IV, fibronectin, laminin-nidogen (12, 13)), gastrointestinal peptides (gastrin-releasing peptide fragment 14-27, gastrin 17 (14)) and cytokines (osteopontin, monocyte chemoattractant protein-1, interleukin-1β (14-16)). From those classical approaches speculative roles for meprin were inferred for the elimination of vasoactive peptides and polypeptide hormones from blood, for tissue remodelling processes, for regulating peristalsis, secretion and growth of intestine or pancreas and in innate immunity.

The introduction of degradomics in 2002 enabled the identification of protease and protease-substrate repertoires on an organism-wide scale by means of proteomic techniques (17). A variety of hitherto unkown substrates for the metzincin metalloendopeptidases ADAM-17 and MMP-14 were found in conditioned media using different cell-based systems (18-20). Alternatively, a complex protein mixture, namely, human plasma, was digested with recombinant MMP-14 in a cell-free system (21). From these studies it became obvious that metalloendopeptidases function as molecular switches in signalling circuits at the cell surface and in the extracellular milieu (22).

Divergent expression patterns were found for both meprin subunits in developing rat intestine during suckling (first two weeks) and weaning (third week) periods (23). Meprinα mRNA exhibited uniform low levels for the first three postnatal weeks whereas meprinβ mRNA showed a biphasic expression with high levels in the first postnatal week followed by low levels during the weaning period. Colostrum and milk proteins are the first substrates intestinal proteases encounter. We therefore used human milk as a complex polysubstrate for meprin in combination with state-of-the-art proteomic techniques. Historically, the protein components of human milk were subdivided into two fractions, the caseins (α-casein, β-casein) and whey. Although the whey fraction consists of a limited set of major proteins (80-90%) such as α-lactalbumin, serum albumin, serotransferrin, secretory immunoglobulin A and lactoferrin, a substantial number of minor proteins are present as well (24, 25). Interestingly, early milk (first week postpartum), called colostrum, is mainly composed of whey (90%) but during the lactation period the casein fraction increases and the whey fraction descreases dramatically (26). Such longitudinal changes in the whey to casein ratio reflect the needs of the infant during lactation; caseins act as a nutritional source of energy, whey proteins mediate antimicrobial, prebiotic, immunomodulatory and growth-stimulating activities (27).

A potential role for meprin in luminal substrate processing of human milk is further supported by the presence of low hydrochloric acid and pepsin levels in gastric juice of newborns and infants (28). Furthermore, serine protease inhibitors ($\alpha_1$-antitrypsin, $\alpha_1$-antichymotrypsin) in human milk may control proteolytic activity in the developing intestine (29). Thus we hypothesized that upon *in vitro* digestion of human milk with rhmeprinβ followed by electrophoretic separation of digests, LC-MS/MS and protein database searching novel putative substrates may be identified.

## 4.3 Methods and materials

### 4.3.1 Human milk

Human milk was obtained from the University Children's Hospital of Berne, Switzerland. Human milk samples were donated by 20 healthy mothers within the first two weeks of lactation and were kept frozen at -20°C until use. Upon thawing on ice a pool of the 20 samples was prepared and a broad-spectrum protease inhibitor cocktail (Complete, EDTA-free, Roche Diagnostics GmbH, Mannheim, Germany) and 0.05% (w/v) sodium azide were immediately added. Human milk was defatted by centrifugation for 20 min at 10,000 × $g$ at 4°C. The solid fat layer was then carefully removed with a spatula. Protein concentration was determined with the BCA™ Protein Assay Kit (Pierce, Rockford, IL, USA).

### 4.3.2 In vitro digestion of human milk

Rhmeprinβ was overexpressed in baculovirus-infected SF9 insect cells as described (10). Defatted human milk was digested with rhmeprinβ for 5 h at 37°C in a buffer containing 50 mM Tris-HCl pH 7.5, 1 mM $MgCl_2$ and a broad-spectrum protease inhibitor cocktail (Complete, EDTA-free, Roche Diagnostics GmbH, Mannheim, Germany) with or without addition of 5 mM of EDTA. Enzyme to substrate ratio was 1:1,000 (w/w). In control experiments rhmeprinβ was omitted from the incubation mixture. The reaction was stopped by addition of 5 mM of EDTA and digests stored at -20°C until use.

### 4.3.3 In vitro digestion of chick tenascin-C

Rhmeprinα and rhmeprinβ were overexpressed in baculovirus-infected insect cells as described (8, 10). Chick tenascin-C (TN-C) was isolated from conditioned media of chick embryo fibroblasts as described (30). The three TN-C splicing variants were digested with rhmeprinα or rhmeprinβ for 5 h and 21.5 h at 37°C in a buffer containing 50 mM Tris-HCl pH 7.5, 1 mM $MgCl_2$ and a broad-spectrum protease inhibitor cocktail (Complete, EDTA-free, Roche Diagnostics GmbH, Mannheim, Germany). Enzyme to substrate ratio was 1:10 (w/w). The reaction was stopped by addition of 5 mM of EDTA and digests stored at -20°C until use.

## 4.3.4 One-dimensional gel electrophoresis

One-dimensional gel electrophoresis was performed according to previously documented methods (31). Separation of human milk and TN-C digests was done in a PROTEAN II xi Cell or in a Mini-PROTEAN 3 Electrophoresis Cell from Bio-Rad Laboratories (Richmond, USA). A 7.5% T, 2.6% C or a 12.5% T, 2.6% C resolving gel was first prepared and allowed to polymerize before overlaying with a 5% T, 2.6% C stacking gel. Protein samples were heated for 5 min in a reducing buffer (20 mM Tris-HCl pH 6.8, 10% (w/v) glycerol, 2% (w/v) SDS, 100 mM DTT) and then centrifuged for 5 min at 16,100 × $g$ prior to loading. Medium-sized gels to be used for subsequent protein identification were stained by colloidal Coomassie blue (32). Other medium-sized gels were stained by silver (33). Mini gels were exclusively produced for immunoblotting. SDS-PAGE Molecular Weight Standards, Broad Range and Precision Plus Protein™ All Blue Standards were purchased from Bio-Rad Laboratories (Richmond, USA).

## 4.3.5 Protein identification by liquid chromatography-based tandem mass spectrometry and protein database searching

Proteins were identified as described with modifications detailed below (34). Gel bands of interest were cut from the gel and sliced into small cubes of 1-2 mm. The gel pieces were placed in 100 µl of deionized Milli-Q water and then washed for 15 min at room temperature once with 100 µl of 50 mM ammonium bicarbonate/ACN (1:1) and once with 100 µl of 50 mM ammonium bicarbonate alone. After dehydration of the gel matrix in 100 µl ACN followed by 5 min in a vacuum centrifuge proteins were digested with 10 µl of sequencing grade trypsin (Promega, Madison, WI, USA) at a concentration of 10 ng/µl overnight at room temperature. After digestion the gel supernatant was transferred into a polypropylene HPLC vial (Agilent Technologies, Waldbrunn, Germany). A volume of 15 µl 20% (v/v) formic acid was used for further peptide extraction from gel pieces by incubation for 15 min at room temperature with occasional shaking. This extract was combined with the first supernatant. 20 µl of protein digest was loaded onto a self-made microbore column (0.15 mm i.d. × 80 mm length) at a flow rate of ~4 µl/min of solvent A (0.1% (v/v) formic acid in water/ACN 98:2). Columns were packed with GROM-SIL 300 Octyl-6 MB, 5 mm, reversed-phase material

(Grom GmbH, Rottenburg-Haiflingen, Germany). Columns were developed by a bi-phasic ACN gradient of 0 to 5% solvent B (0.1% (v/v) formic acid in water/ACN 4.9:95) in 1 min followed by 5 to 40% solvent B in 20 min at a flow rate of approximately 3 ml/min. The column effluent was directly coupled to an Esquire3,000+ ion trap mass spectrometer from Bruker Daltonics (Bremen, Germany) via a capillary ESI source operated at 3,700 V. CID was triggered on the two most abundant not singly charged peptide ions in the $m/z$ range of 360-1,400. Precursors were set in an exclusion list for 0.5 min. CID spectra interpretation was performed with Phenyx on vital-it.ch server operated by GeneBio (Geneva, Switzerland) against the uniprot-SwissProt protein database (release 51.4) with variable oxidation of methionine and variable deamidation of asparagine and glutamine. Parent and fragment mass tolerances were set to 0.8 Da. Up to two missed cleavages and half tryptic peptides were allowed. The taxonomic search space was restricted to Mammalia. All protein identifications consisting of at least 2 peptides reaching a p-value of ≤0.00001 were accepted.

### *4.3.6 Functional classification*

Proteins were clustered into functional groups according to the Human Protein Reference Database (http://hprd.ord/) (35).

### *4.3.7 Immunoblotting*

Western blotting was performed as described (36). Briefly, following one-dimensional gel electrophoresis as detailed above the protein was transferred to a Hybond™-P PVDF membrane (Amersham Biosciences, Uppsala, Sweden) by application of a constant potential for 15 min at 30 V and 50 min at 80 V. Following protein transfer the membrane was incubated overnight at 4°C in blocking solution TTBS (20 mM Tris-HCl pH 7.5, 137 mM NaCl, 0.1% (w/v) Tween-20) containing 5% (w/v) milk powder. The membrane was washed twice in TTBS for 1 min and 10 min followed by incubation for 1 h with primary antibody prepared in antibody solution TTBS containing 2% (w/v) milk powder. Membrane was washed four times in TTBS for 1 min, 15 min and 5 min (twice). The secondary antibody was horseradish peroxidase-linked sheep anti-mouse (Amersham Biosciences, Uppsala, Sweden) diluted 1:10,000 in antibody solution. Membrane was incubated for 1 h followed by washing four times in TTBS for 1 min, 15 min and 5 min (twice). The secondary antibody was visualized using the ECL plus Western Blotting Detection System (Amersham Biosciences,

Uppsala, Sweden). Monoclonal antibodies against specific domains of chick TN-C were essentially used as described: anti-Tn60 (30, 37) diluted 1:1,000, anti-TnM1 (30, 37-39) diluted 1:100, anti-Tn20 (39) diluted 1:16,000, anti-Tn26 (30, 37) diluted 1:16,000, anti-Tn32 (30, 37) diluted 1:16,000, anti-Tn68 (30, 37-39) diluted 1:4,000 and finally, anti-Tn4 (39) diluted 1:1,000.

## 4.4 Results

### 4.4.1 In vitro digestion of human milk by recombinant human meprin beta

The protein concentration in human milk changes longitudinally from colostrum (first week postpartum) through transitional to mature milk (one to three months postpartum) (26). Furthermore, the protein composition of human milk varies diurnally and individually. Thus we decided to combine human milk samples from 20 different mothers. All samples were collected within the first two weeks of lactation. The pooled human milk was then defatted after addition of a broad-spectrum protease inhibitor cocktail to avoid unwanted intrinsic proteolytic activity. In preliminary experiments defatted human milk was incubated with rhmeprinβ at varying enzyme to substrate molar ratios of 1:10, 1:100, 1:500 and 1:1000 (data not shown). Analysis of these milk digests by one-dimensional gel electrophoresis showed consistent substrate cleavage patterns common to all enzyme to substrate ratios tested. Three major groups of electrophoretically separated milk proteins running at approximate molecular weight positions of 200 kDa (bands A-D), 31 kDa (bands F-H) and 21 kDa (bands I and J), respectively, were subject to proteolysis by rhmeprinβ (Fig. 1). In order to identify the substrate degradome of rhmeprinβ in human milk gel bands of interest were then prepared for subsequent protein identification as detailed below.

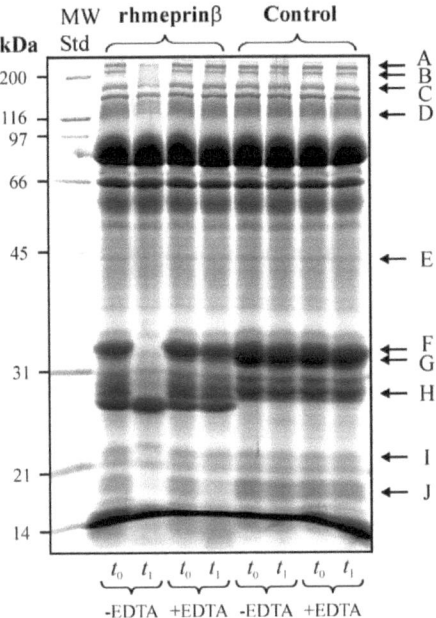

**Fig. 1. The substrate degradome of recombinant human meprin beta in human milk.**
Defatted human milk was incubated with rhmeprinβ for 5 h at 37°C. The incubation mixture contained a broad-spectrum protease inhibitor cocktail that was EDTA-free. Metalloendopeptidase activity was inhibited by addition of 5 mM of EDTA in conditions indicated. Enzyme to substrate ratio was 1:1000 (w/w). Control experiments were performed without rhmeprinβ. One-dimensional separation of human milk digests (150 µg per lane) in a 12.5% SDS gel under reducing conditions. Colloidal Coomassie blue staining. Migration positions of molecular weight standards (MW Std) are shown on the gel. Gel bands of interest were cut and sliced for subsequent in-gel tryptic digestion, LC-MS/MS and protein database searching (Table 1). The band is identified with a letter.

***4.4.2 Functional classification of human milk proteins identified by liquid chromatography-based tandem mass spectrometry and protein database searching***

The rapidly growing field of proteomics offers a diverse set of very powerful microsequencing platforms to characterize unknown proteins (17). Hence we applied in-gel tryptic digestion, LC-MS/MS and protein database searching to identify human milk proteins prone to proteolysis by rhmeprinβ. Collision-induced dissociation spectra interpretation with the Phenyx public search engine against the uniprot-SwissProt protein database (release 51.4) allowed for 16 unambiguous protein identifications (Fig. 1 and Table 1). The taxonomic search space was restricted to Mammalia. All peptide sequence tags obtained were from the human. In addition, multiple unique proteins were found in gel bands B, E and J, respectively.

The prime goal of systems biology is to define general rules deduced from the outcome of a system's response upon its perturbation. In the context of degradomics this may postulate to cluster identified substrates into decisive categories. Therefore proteins identified by LC-MS/MS were classified into functional groups according to the Human Protein Reference Database (35) (Table 1). Interestingly, six proteins were related to the biological process "transport", four to "cell communication/signal transduction" and two to "metabolism/energy pathways". The four remaining proteins could not be functionally classified. Thus hmeprinβ may play a regulatory role in the processing of proteins involved in transport of metal ions, in extracellular signalling and in lipid metabolism. Our findings match the current opinion that metalloendopeptidases are considered to be regulatory scissors rather than ECM degrading bulldozers (22).

We clearly demonstrated the applicability of proteomics to protease substrate discovery, nonetheless novel cleavage candidates need to be verified by means of alternative techniques. Due to the complexity of data generated follow up validation experiments may be restricted to only a few putative protease targets. Human TN-C was found in two distinct bands, namely, gel band A and B (Fig. 1 and Table 1), and hence selected for *in vitro* cleavage assays as shown below.

**Table 1**

**Protein identification by liquid chromatography-based tandem mass spectrometry and functional classification**

In-gel tryptic digestion of human milk proteins from gel bands of interest. Peptide separation/fragmentation with LC-MS/MS. CID spectra interpretation with Phenyx on vital-it.ch against the uniprot-SwissProt protein database (release 51.4). The band is identified with a letter. All proteins identified were from the human. The protein score is the sum of the best peptide z-scores per valid peptide sequences. Minimum peptide z-score restricted to 5. Theoretical mass of a given uniprot-SwissProt protein database entry calculated with the ExPASy "Compute pI/Mw" tool (http://www.expasy.org/tools/pi_tool.html). Functional classification according to Human Protein Reference Database (35).

| ID | Protein identification | Protein description | Swiss-Prot accession Number | Number of unique peptides | Protein score | Theoretical mass (Da) | Biological process |
|---|---|---|---|---|---|---|---|
| A | TENA | Tenascin-C | P24821 | 20 | 196.94 | 238 659 | Cell communication; Signal transduction |
| B | FAS | Fatty acid synthase | P49327 | 8 | 73.44 | 273 400 | Metabolism; Energy pathways |
|  | TENA | Tenascin-C | P24821 | 4 | 36.23 | 238 659 | Cell communication; Signal transduction |
| C | MRC1 | Macrophage mannose receptor 1 | P22897 | 4 | 41.15 | 164 120 | Cell communication; Signal transduction |
| D | CEL | Bile salt-activated lipase | P19835 | 7 | 60.92 | 76 272 | Metabolism; Energy pathways |
| E | ACTB | Beta-actin | P60709 | 8 | 76.31 | 41 606 | - |
|  | CASK | Kappa-casein | P07498 | 7 | 57.13 | 18 163 | Biological_process unknown |
|  | MFGM | Milk fat globule-EGF factor 8 | Q08431 | 2 | 19.46 | 40 791 | Cell communication; Signal transduction |
| F | CASB | Beta-casein | P05814 | 2 | 17.34 | 23 858 | Transport |
| G | CASB | Beta-casein | P05814 | 2 | 16.99 | 23 858 | Transport |
| H | CASB | Beta-casein | P05814 | 2 | 16.8 | 23 858 | Transport |
| I | CASB | Beta-casein | P05814 | 2 | 19.87 | 23 858 | Transport |
| J | CASA1 | Alpha-S1-casein | P47710 | 3 | 24.36 | 20 089 | Transport |
|  | CASB | Beta-casein | P05814 | 2 | 18.51 | 23 858 | Transport |
|  | PIP | Prolactin-inducible protein | P12273 | 2 | 23 | 13 523 | Biological_process unkown |
|  | PPIA | Cyclosporin A-binding protein | P62937 | 2 | 16.18 | 17 881 | - |

### 4.4.3 In vitro cleavage of purified chick tenascin-C splicing variants by recombinant human meprin alpha and beta

The two forms of TN-C that we identified in human milk may represent different splicing variants originating from the same gene. Human TN-C exhibits nine variable fibronectin type III (FN-III) repeats subjected to alternative splicing (40). In contrast, chick TN-C contains only three alternative FN-III repeats dramatically reducing the combinatorial complexity of existing splicing variants (37). Chick TN-C isolated from conditioned media of chick embryo fibroblasts features three different forms only: the 230 kDa variant (Tn230), the 200 kDa variant (Tn200) and the 190 kDa variant (Tn190) (Fig. 2A) (30). Thus we used chick TN-C as simplified model system to test susceptibility of individual splicing variants to hydrolysis by both meprin subunits. Marked differences between the two mouse meprin orthologs in substrate and peptide bond specificity had been previously reported (14).

Purified chick TN-C was digested with rhmeprin$\alpha$ or $\beta$ at an enzyme to substrate molar ratio of 1:10 for 5 h and 21.5 h, respectively. Unspecific (non-meprin) proteolytic activity was avoided by a broad-spectrum protease inhibitor cocktail added to the incubation mixture. The TN-C digests and controls were then separated by one-dimensional gel electrophoresis followed by post-electrophoretic visualization with silver staining (Fig. 3). In controls two major bands with molecular weights of 70 kDa and 64 kD were detected for rhmeprin$\alpha$ whereas rhmeprin$\beta$ displayed a single band of 68 kDa. Upon prolonged incubation the 70 kDa and 64 kDa bands related to rhmeprin$\alpha$ were marginally degraded to 62 kDa and 56 kDa fragments probably due to autolysis. Rhmeprin$\alpha$ equally processed all three TN-C variants into 220 kDa, 190 kDa and 180 kDa forms presumably by cleaving off a 10 kDa peptide. Rhmeprin$\beta$ preferably cleaved the three TN-C variants into 180 kDa, 150 kDa and 40 kDa fragments. Interestingly, short-term incubation with rhmeprin$\beta$ revealed transient degradation products running at molecular molecular weight positions of 200 kDa, 190 kDa and 180 kDa. Furthermore, upon prolonged incubation rhmeprin$\alpha$ was able to produce a 40 kDa fragment as well.

Therefore we hypothesized that both meprin subunits may cleave off 10 kDa and 40 kDa peptides irrespective of available alternative FN-III repeats, but rhmeprin$\beta$ may be more effective in removing the longer peptide fragments and rhmeprin$\alpha$ in elminating the shorter ones. Next, to figure out the origin of those cleavage products we performed immunoblotting experiments with domain-specific monoclonal antibodies as detailed below.

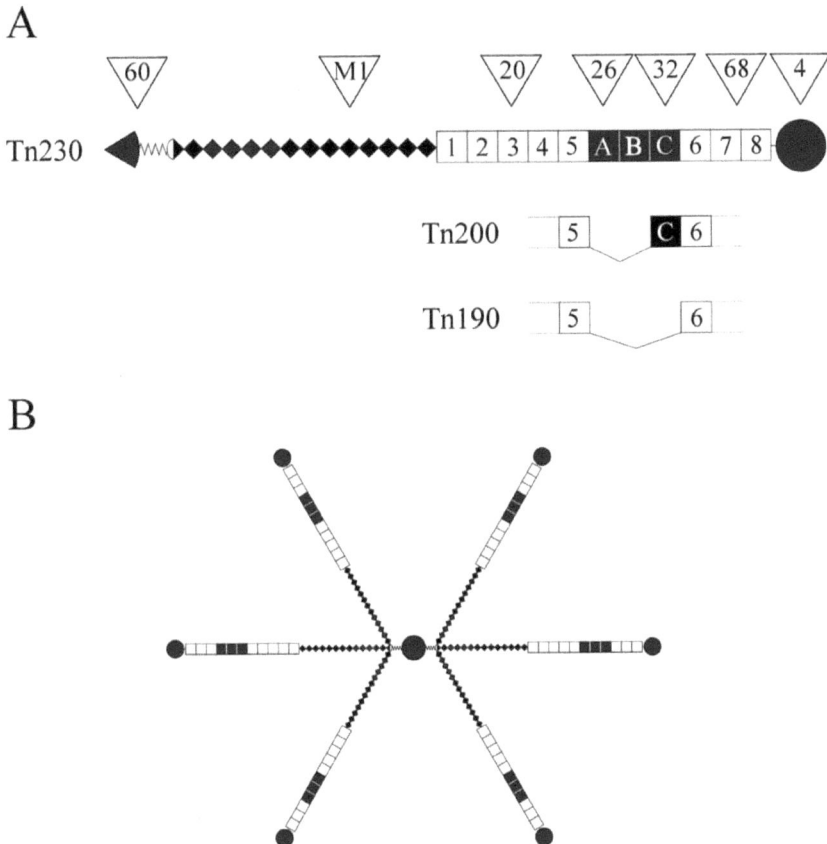

**Fig. 2. Chick tenascin-C: multidomain organization, splicing variants, oligomeric state and epitope localization of domain-specific monoclonal antibodies.**

(A) Distinct protein domains were depicted with specific symbols. At the N-terminus a sector of a circle represents the part contributing to the central globule and a wavy line the heptad repeats. After another circle segment 13.5 EGF-like repeats are indicated as diamonds. The boxes correspond to FN-III repeats. A short dash leads to the C-terminal globule homologous to fibrinogen. Omission of the FN-III repeats A and B, which are specific for the 230 kDa splicing variant Tn230, leads to the middle tenascin form indicated as Tn200. Further splicing out of the C repeat results in the shortest version Tn190. Numbered open triangles mark the epitopes of the corresponding domain-specific monoclonal antibodies as described: anti-Tn60, anti-TnM1, anti-Tn20, anti-Tn26, anti-Tn32, anti-Tn68 and anti-Tn4.

(B) Each TN-C splicing variant is able to build up a homohexameric (hexabrachion) structure that is composed of two triplets of arms joined in the triple coiled coil region via amphipathic α-helices of heptad repeats. For simplicity, the hexabrachion structure of Tn230 is shown.

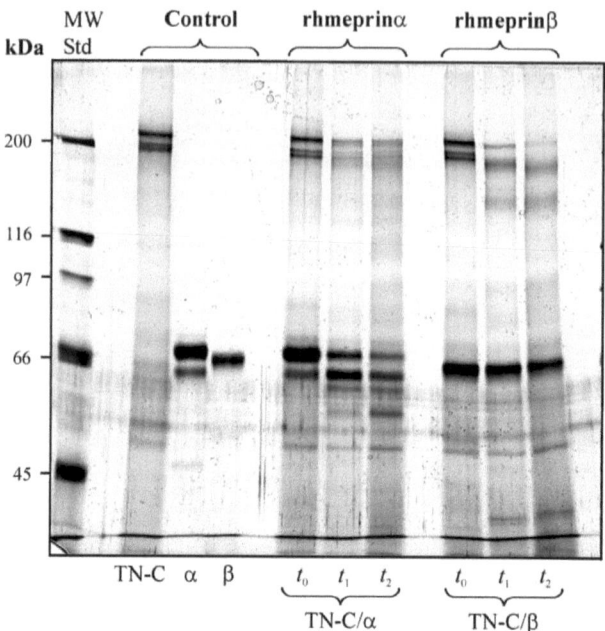

**Fig. 3.** *In vitro* **cleavage of purified chick tenascin-C splicing variants by recombinant human meprin alpha and beta.**

The three TN-C splicing variants from conditioned media of chick embryo fibroblasts were incubated with rhmeprinα or rhmeprinβ for 5 h and 21.5 h at 37°C. The incubation mixture contained a broad-spectrum protease inhibitor cocktail that was EDTA-free. Enzyme to substrate ratio was 1:10 (w/w). One-dimensional separation of TN-C digests (6 µg per lane) in a 7.5% SDS gel under reducing conditions. Controls were loaded at same molar ratios as used in cleavage assay. Silver staining. Migration positions of molecular weight standards (MW Std) are shown on the gel.

## 4.4.4 Immunoblotting of recombinant human meprin alpha- and beta-generated chick tenascin-C digests with domain-specific monoclonal antibodies

Chick TN-C comprises a multidomain protein architecture; namely, a N-terminal central globule, heptad repeats, another circle segment, 13.5 EGF-like repeats, eight constant and three variable FN-III repeats and a C-terminal fibrinogen-like globule (Fig. 2A) (37). A variety of distinct domain-specific monoclonal antibodies against chick TN-C were previously described: anti-Tn60 (30, 37), anti-TnM1 (30, 37-39), anti-Tn20 (39), anti-Tn26 (30, 37), anti-Tn32 (30, 37), anti-Tn68 (30, 37-39) and anti-Tn4 (39). Thus the availability of a complete set of monoclonal antibodies enables specific mapping of cleavage sites on rhmeprinα- and rhmeprinβ-generated chick TN-C fragments (Fig. 2A).

*In vitro* digestion of purified chick TN-C by rhmeprinα or β and one-dimensional gel electrophoresis were done as detailed above. Subsequently, westernblotting experiments were performed with different domain-specific monoclonal antibodies to detect individual cleavage patterns (Fig. 4).

Anti-Tn4 (Fig. 4A): Both meprin subunits generated a C-terminal 40 kDa fragment. Furthermore, anti-Tn4 recognized the 220 kDa, 190 kDa and most likely also the 180 kDa fragments obtained with rhmeprinα incubation. Hence the putative 10 kDa peptide mentioned above was definitively removed from the N-terminus. Moreover, rhmeprinβ efficiently processed all long C-terminal fragments (220 kDa, 190 kDa and 180 kDa) as these disappeared. Therefore both meprin subunits may process all three TN-C splicing variants, but rhmeprinα preferably by removal of the N-terminal 10 kDa peptide and rhmeprinβ by cleaving off the C-terminal 40 kDa peptide.

Anti-Tn68 (Fig. 4B): The C-terminal 40 kDa peptide produced by both meprin subunits was absent. Furthermore, the expected rhmeprinβ-generated 180 kDa and 150 kDa cleavage products could not be detected as well. Conclusively, the C-terminal cleavage site is located exactly within the epitope of anti-Tn68 in the seventh FN-III repeat. The long C-terminal fragments (220 kDa, 190 kDa and 180 kDa) in rhmeprinα digests may represent N-terminally truncated versions derived from all three TN-C splicing variants that were not cleaved yet within the seventh FN-III repeat.

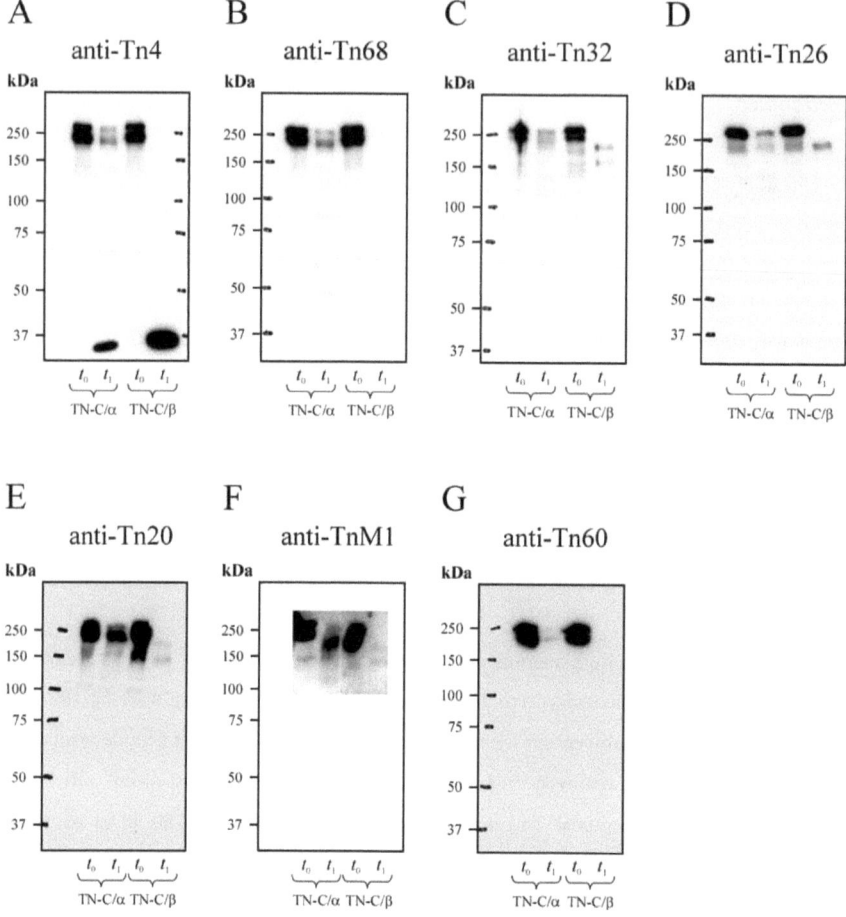

Fig. 4. Mapping the cleavage sites on recombinant human meprin alpha- and beta-generated chick tenascin-C fragments with domain-specific monoclonal antibodies.

The three TN-C splicing variants from conditioned media of chick embryo fibroblasts were incubated with rhmeprinα or rhmeprinβ for 21.5 h at 37°C. The incubation mixture contained a broad-spectrum protease inhibitor cocktail that was EDTA-free. Enzyme to substrate ratio was 1:10 (w/w). One-dimensional separation of TN-C digests in a 7.5% SDS gel under reducing conditions. Western blotting was performed with domain-specific monoclonal antibodies as follows: (A) anti-Tn4, (B) anti-Tn68, (C) anti-Tn32, (D) anti-Tn26, (E) anti-Tn20, (F) anti-TnM1 and (G) anti-Tn60. Migration positions of molecular weight standards are shown on the blot.

Anti-Tn32 (Fig. 4C): This antibody specifically binds to the Tn230 and Tn200 variants (Fig. 2A). The 180 kDa and 150 kDa fragments could be recognized, thus they may constitute N-terminally shortened fragments having lost first the 10 kDa and thereafter the C-terminal 40 kDa peptide. In conclusion, rhmeprin$\alpha$ preferably attacks the N-terminus whereas rhmeprin$\beta$ may truncate both ends of TN-C at same efficiency.

Anti-Tn26 (Fig. 4D): This antibody singly recognizes the Tn230 form (Fig. 2A). Hence the 180 kDa cleavage product definitively originates from the longest TN-C splicing variant. Therefore the 150 kDa fragment was derived from the Tn200 variant.

Anti-Tn20 (Fig. 4E): This antibody detects all three TN-C versions (Fig. 2A). As expected the two N-terminal 180 kDa and 150 kDa fragments were both detectable.

Anti-TnM1 (Fig. 4F): This antibody was known not to optimally react under reducing conditions. Nevertheless same cleavage patterns were demonstrated as seen with anti-Tn20.

Anti-Tn60 (Fig. 4G): The epitope of this antibody is located within the N-terminal region (Fig. 2A). Removal of the N-terminal 10 kDa peptide may disrupt the binding capacity of anti-Tn60. Thus the putative fragments observed upon digestion of TN-C by rhmeprin$\alpha$ may represent remaining forms of intact subunits, namely, 230 kDa, 220 kDa and 190 kDa in size.

In summary, we successfully applied a broad array of domain-specific monoclonal antibodies to accurately map the cleavage sites within the diverse chick TN-C splicing variants prone to hydrolysis by rhmeprin$\alpha$ and $\beta$. The two meprin forms showed distinct cleavage preferences for chick TN-C; the $\beta$ subunit equally cleaved off the N- and C-terminal peptides whereas the $\alpha$ subunit preferentially removed the 10 kDa peptide (Fig. 5A). Moreover, the differentially spliced FN-III domains of TN-C may not determine substrate cleavage specificity for the two meprin subunits.

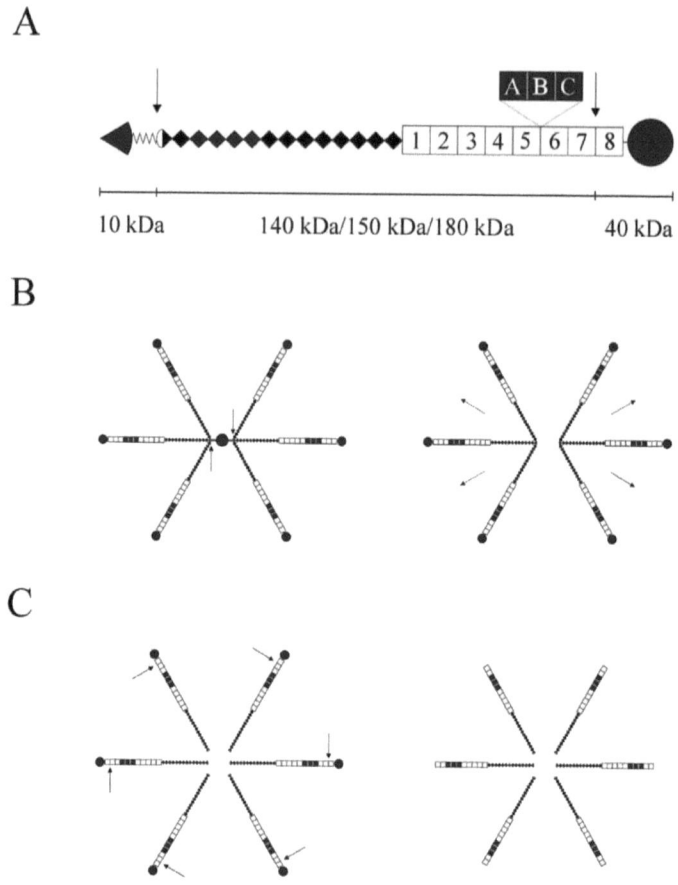

**Fig. 5. Model of chick tenascin-C cleavage sites targeted by recombinant human meprin alpha or beta and putative cleavage mechanism.**

(A) The multidomain organization of chick TN-C is represented by specific symbols as detailed in Fig. 2. For simplicity, the shortest splicing variant Tn190 is shown with 8 constant FN-III repeats. Alternative FN-III repeats A-C are depicted as an insert. Solid arrows point to the N- and the C-terminal cleavage sites of the three TN-C splicing variants targeted by rhmeprinα and rhmeprinβ. The 10 kDa N-terminal fragment and the 40 kDa C-terminal fragment originate from all TN-C splicing variants. Cleavage at both ends generates a 140 kDa fragment for Tn190, a 150 kDa fragment for Tn200 and a 180 kDa fragment for Tn230.

(B) The N-terminal cleavage (solid arrows) may disrupt the homohexameric TN-C structure into two trimers that finally fall apart (dotted arrows) into monomeric TN-C. For simplicity, the Tn230 hexabrachion is depicted.

(C) The C-terminal cleavage (solid arrows) of TN-C within its cell-binding site (residues 1406-1509) may lead to detachment from cells or other substrates. For simplicity, the Tn230 hexabrachion is depicted.

## 4.5 Discussion

### 4.5.1 Meprin beta and its substrate repertoire in human milk

In recent years degradomics was applied to decipher the substrate repertoires of metalloendopeptidases on a system-wide level (17-22). Due to the rapid development of proteomic techniques two methodolocial platforms were successfully used for protein separation; liquid chromatography and two-dimensional gel electrophoresis (18-21). Another high-priority issue was the selection of a complex protein mixture, namely, culture medium or human plasma, in an appropriate biological context to find physiologically relevant protease substrates (18-21).

We decided to use human milk as a complex polysubstrate to elucidate novel putative roles for meprin. Human milk is a complex protein mixture known to change its protein composition diurnally, longitudinally and individually during lactation (26). Thus the prime goal of this project was to design a simple *in vitro* cell-free system respecting variability in milk protein composition (Fig. 1). In a previous report meprinβ but not α mRNA was demonstrated to be predominantly expressed in developing rat intestine during suckling period (first two weeks), hence colostrum (breast milk of first two weeks postpartum) represented an appropriate polysubstrate to be treated with rhmeprinβ (23). Furthermore, low gastric acid secretion and low pepsin concentration in gastric juice of neonates and finally, presence of serine protease inhibitors ($\alpha_1$-antitrypsin, $\alpha_1$-antichymotrypsin) in human milk supported a role for meprin in luminal substrate processing of milk proteins (28, 29).

Interestingly, upon functional classification of novel substrates cleaved by rhmeprinβ three groups were identified: "transport", "metabolism/energy pathways" and "cell communication/signal transduction" (Table 1). Hence implicating novel potential roles for meprin in the processing of milk proteins exerting functions such as metal tranport, lipid metabolism and extracellular signalling.

Recently, it was realized that metalloendopeptidases act as molecular switches in signalling circuits at the cell surface and in the extracellular milieu, therefore meprin may process its substrates by targeted cleavage of scissile bonds leading to degradation products with impaired or novel biological functions (22).

### 4.5.2 Human tenascin-C a novel component in human milk

The major protein components of human milk were identified by two-dimensional gel electrophoresis and microsequencing techniques; namely, α- and β-casein families, α-lactalbumin, serum albumin, serotransferrin, secretory immunoglobulin A, lactoferrin and numerous minor proteins (24, 25). Interestingly, we found a novel hitherto unkown protein component, TN-C, in human milk (Fig. 1 and Table 1). TN-C is an adhesion-modulating ECM protein predominantly expressed during development and tissue remodelling processes (40). In a previous report TN-C mRNA was detected in epithelial cells of the lactating murine mammary gland and on the protein level in murine milk within the first week of lactation (41). The authors did not offer an explanation for the significance of secreted TN-C into milk. The presence of TN-C in milk early during lactation may be explained by the fact that TN-C expression is stronlgy up-regulated by transforming growth factor-β, another constituent of human milk (42, 43). Moreover, the concentration of hormones and growth factors in milk is highest in colostrum (first two weeks postpartum) (44). In addition, TN-C strongly inhibits β-casein protein synthesis in lactating murine mammary epithelial cells (45). The casein fraction of colostral milk is very low (10%) early in lactation (first week postpartum) but increases dramatically during the lactation period (26). Thus we may speculate that TN-C present in colostrum regulates protein synthesis in transition from early to mature milk; the biological relevance of this finding needs further investigation.

### 4.5.3 Distinct functions for the two meprin subunits alpha and beta exemplified by tenascin-C proteolytic processing

Human TN-C was identified in two different bands that were prone to proteolysis by rhmeprinβ (Fig. 1 and Table 1). The two forms may originate from differential splicing within the nine potential alternative FN-III repeats featured by human TN-C (40). In contrast, chick TN-C comprises only three alternative splicing variants (Fig. 2A) (30). Hence a simple chick TN-C model system was used to dissect the individual cleavage preferences for the two meprin subunits α and β in substrate processing. *In vitro* digestion of all three chick TN-C variants by rhmeprinα and rhmeprinβ clearly demonstrated distinct functions for the two meprin subunits; meprinα preferentially removed 10 kDa peptides producing 220 kDa, 190 kDa and probably also 180 kDa cleavage products whereas meprinβ cleaved off 10 kDa and

40 kDa peptides leading to 180 kDa and 150 kDa fragments (Fig. 3). We then successfully mapped the cleavage sites on rhmeprinα- and rhmeprinβ-generated chick TN-C fragments with a complete set of domain-specific monoclonal antibodies (Fig. 2A and Fig. 4). This mapping strategy finally revealed that both meprin subunits equally processed all three TN-C variants but not by selecting the two putative cleavages sites at same efficiency; meprinα preferably cut off N-terminally-derived 10 kDa peptides whereas meprinβ removed probably first these N-terminal peptides followed by a second C-terminal cleavage eliminating 40 kDa peptides (Fig. 4 and Fig. 5A). Furthermore, the C-terminal cleavage of all three TN-C variants occurred exactly within the seventh FN-III repeat, the cell-binding site (Fig. 4 and Fig. 5) (37).

Ultrastructurally, human and chick TN-C exhibit a hexabrachion structure consisting of six tenascin monomers (Fig. 2B) (37). A N-terminal cleavage within the epitope region of anti-TN60 disrupts the structural integrity of the tenascin hexamer resulting in monomeric forms (Fig. 5B). This putative cleavage mechanism was previously hypothesized for the matrix metalloproteinase MMP7 which was shown to evenly process both long and small human TN-C isoforms by removal of the amino-terminal knob (46, 47).

Conversely, the C-terminal proteolytic degradation of TN-C within its cell-binding site may lead to detachment from cells and other binding partners (Fig. 5C). In a previous study it was reported that mild tryptic treatment of chick TN-C led to removal of a 10 kDa peptide from the C-terminal fibrinogen-like globule of the smallest splicing variant (Tn190), abolishing its ability to bind to heparin-agarose (39). More recently, it was demonstrated that binding of human TN-C to the transmembrane heparan sulfate proteoglycan Syndecan-4 blocked fibronectin-mediated cell adhesion (48). Moreover, TN-C concomitantly bound to Syndecan-4 and fibronectin via its fibrinogen-like domain and C-terminal FN-III domains, respectively (48). Therefore a potential cleavage within the C-terminal FN-III repeat of TN-C preferably executed by meprinβ may counteract interference with fibronectin-mediated cell spreading. Nevertheless this hypothesis needs to be verified by cell culture system-based experiments.

In conclusion, these findings may point to distinct functions for the two meprin subunits α and β exemplified by different cleavage preference in proteolytic processing of TN-C. Marked differences between α- and β-subunits in substrate and peptide bond specificity were previously reported; meprinα selects for small (e.g. serine, alanine and threonine) or hydrophobic (e.g. phenylalanine) residues in P1 and P1' sites and proline in P2' position whereas meprinβ prefers acidic amino acids in the P1 and P1' sites and selects against basic residues at P2' and P3' (14). Here we clearly demonstrate that all three chick TN-C splicing variants are processed by both meprin subunits at same efficiency but at distinct sites. Hence the extra FN-III repeats may not determine distinct cleavage preference for the α- an β-subunits. Interestingly, these findings may not reflect the current knowledge of TN-C proteolytic processing executed by other metalloendopeptidases. The matrix metalloproteinases MMP-2 and MMP-3 exclusively cleave long human TN-C isoforms within the alternatively spliced region encompassing a single extra FN-III repeat (46). Furthermore, MMP-7 cleaves long human TN-C isoforms within three extra FN-III repeats and MMP-9 is not able to cleave any TN-C splicing variant (46). The astacin-like metalloendopeptidases meprinα and meprinβ thus represent novel unique scissors that may exert highly important regulatory processing of TN-C during development and tissue remodelling processes. This conclusion complies with the current belief that metalloendopeptidases are regulatory scissors modulating intra- and extracellular signalling events (22).

## 4.6 References

1. Sterchi EE, Green JR, Lentze MJ. Non-pancreatic hydrolysis of N-benzoyl-l-tyrosyl-p-aminobenzoic acid (PABA-peptide) in the human small intestine. *Clin Sci* 1982;62:557-60.
2. Sterchi EE, Naim HY, Lentze MJ, Hauri HP, Fransen JA. N-benzoyl-L-tyrosyl-p-aminobenzoic acid hydrolase: a metalloendopeptidase of the human intestinal microvillus membrane which degrades biologically active peptides. *Arch Biochem Biophys* 1988;265:105-18.
3. Beynon RJ, Shannon JD, Bond JS. Purification and characterization of a metallo-endoproteinase from mouse kidney. *Biochem J* 1981;199:591-8.
4. Kenny AJ, Fulcher IS, Ridgwell K, Ingram J. Microvillar membrane neutral endopeptidases. *Acta Biol Med Ger* 1981;40:1465-71.
5. Grünberg J, Dumermuth E, Eldering JA, Sterchi EE. Expression of the alpha subunit of PABA peptide hydrolase (EC 3.4.24.18) in MDCK cells. Synthesis and secretion of an enzymatically inactive homodimer. *FEBS Lett* 1993;335:376-9.
6. Eldering JA, Grünberg J, Hahn D, Croes HJ, Fransen JA, Sterchi EE. Polarised expression of human intestinal N-benzoyl-L-tyrosyl-p-aminobenzoic acid hydrolase (human meprin) alpha and beta subunits in Madin-Darby canine kidney cells. *Eur J Biochem* 1997;247:920-32.
7. Rösmann S, Hahn D, Lottaz D, Kruse MN, Stöcker W, Sterchi EE. Activation of human-meprin alpha in a cell culture model of colorectal cancer is triggered by the plasminogen-activating system. *J Biol Chem* 2002;277:40650-8.
8. Becker C, Kruse MN, Slotty KA, Köhler D, Harris JR, Rösmann S, et al. Differences in the activation mechanism between the alpha and beta subunits of human meprin. *Biol Chem* 2003;384:825-31.
9. Becker-Pauly C, Howel M, Walker T, Vlad A, Aufenvenne K, Oji V, et al. The alpha and beta subunits of the metalloprotease meprin are expressed in separate layers of human epidermis, revealing different functions in keratinocyte proliferation and differentiation. *J Invest Dermatol* 2006;[Epub ahead of print].
10. Köhler D, Kruse MN, Stöcker W, Sterchi EE. Heterologously overexpressed, affinity-purified human meprin alpha is functionally active and cleaves components of the basement membrane in vitro. *FEBS Lett* 2000;465:2-7.
11. Yamaguchi T, Fukase M, Kido H, Sugimoto T, Katunuma N, Chihara K. Meprin is predominantly involved in parathyroid hormone degradation by the microvillar membranes of rat kidney. *Life Sci* 1994;54:381-6.
12. Kaushal GP, Walker PD, Shah SV. An old enzyme with a new function: purification and characterization of a distinct matrix-degrading metalloproteinase in rat kidney cortex and its identification as meprin. *J Cell Biol* 1994;126:1319-27.
13. Walker PD, Kaushal GP, Shah SV. Meprin A, the major matrix degrading enzyme in renal tubules, produces a novel nidogen fragment in vitro and in vivo. *Kidney Int* 1998;53:1673-80.
14. Bertenshaw GP, Turk BE, Hubbard SJ, Matters GL, Bylander JE, Crisman JM, et al. Marked differences between metalloproteases meprin A and B in substrate and peptide bond specificity. *J Biol Chem* 2001;276:13248-55.
15. Villa JP, Bertenshaw GP, Bylander JE, Bond JS. Meprin proteolytic complexes at the cell surface and in extracellular spaces. *Biochem Soc Symp* 2003;53-63.

16. Herzog C, Kaushal GP, Haun RS. Generation of biologically active interleukin-1beta by meprin B. *Cytokine* 2005;31:394-403.
17. Lopez-Otin C, Overall CM. Protease degradomics: a new challenge for proteomics. *Nat Rev Mol Cell Biol* 2002;3:509-19.
18. Guo L, Eisenman JR, Mahimkar RM, Peschon JJ, Paxton RJ, Black RA, et al. A proteomic approach for the identification of cell-surface proteins shed by metalloproteases. *Mol Cell Proteomics* 2002;1:30-36.
19. Tam EM, Morrison CJ, Wu YI, Stack MS, Overall CM. Membrane protease proteomics: Isotope-coded affinity tag MS identification of undescribed MT1-matrix metalloproteinase substrates. *Proc Natl Acad Sci USA* 2004;101:6917-22.
20. Bech-Serra JJ, Santiago-Josefat B, Esselens C, Saftig P, Baselga J, Arribas J, et al. Proteomic identification of desmoglein 2 and activated leukocyte cell adhesion molecule as substrates of ADAM17 and ADAM10 by difference gel electrophoresis. *Mol Cell Biol* 2006;26:5086-95.
21. Hwang IK, Park SM, Kim SY, Lee ST. A proteomic approach to identify substrates of matrix metalloproteinase-14 in human plasma. *Biochim Biophys Acta* 2004;1702:79-87.
22. Overall CM, Blobel CP. In search of partners: linking extracellular proteases to substrates. *Nat Rev Mol Cell Biol* 2007;8:245-57.
23. Henning SJ, Oesterreicher TJ, Osterholm DE, Lottaz D, Hahn D, Sterchi EE. Meprin mRNA in rat intestine during normal and glucocorticoid-induced maturation: divergent patterns of expression of alpha and beta subunits. *FEBS Lett* 1999;462:368-72.
24. Anderson NG, Powers MT, Tollaksen SL. Proteins of human milk. I. Identification of major components. *Clin Chem* 1982;28:1045-55.
25. Murakami K, Lagarde M, Yuki Y. Identification of minor proteins of human colostrum and mature milk by two-dimensional electrophoresis. *Electrophoresis* 1998;19:2521-7.
26. Lönnerdal B. Effects of maternal dietary intake on human milk composition. *J Nutr* 1986;116:499-513.
27. Lönnerdal B. Nutritional and physiologic significance of human milk proteins. *Am J Clin Nutr* 2003;77:1537S-43S.
28. Agunod M, Yamaguchi N, Lopez R, Luhby AL, Glass GB. Correlative study of hydrochloric acid, pepsin, and intrinsic factor secretion in newborns and infants. *Am J Dig Dis* 1969;14:400-14.
29. Chowanadisai W, Lönnerdal B. Alpha (1)-antitrypsin and antichymotrypsin in human milk: origin, concentrations, and stability. *Am J Clin Nutr* 2002;76:828-33.
30. Chiquet M, Vrucinic-Filipi N, Schenk S, Beck K, Chiquet-Ehrismann R. Isolation of chick tenascin variants and fragments. A C-terminal heparin-binding fragment produced by cleavage of the extra domain from the largest subunit splicing variant. *Eur J Biochem* 1991;199:379-88.
31. Laemmli UK. Cleavage of structural proteins during the assembly of the head of bacteriophage T4. *Nature* 1970;15:680-85.
32. Anderson NL, Esquer-Blasco R, Hofmann JP, Anderson NG. A two-dimensional gel database of rat liver proteins useful in gene regulation and drug effect studies. *Electrophoresis* 1991;12:907-30.
33. Swain M, Ross NW. A silver stain protocol for proteins yielding high resolution and transparent background in sodium dodecyl sulfate-polyacrylamide gels. *Electrophoresis* 1995;16:948-51.

34. Heller M, Stalder D, Schlappritzi E, Hayn G, Matter U, Haeberli A. Mass spectrometry-based analytical tools for the molecular protein characterization of human plasma lipoproteins. *Proteomics* 2005;5:2619-30.
35. Peri S, Navarro JD, Amanchy R, Kristiansen TZ, Jonnalagadda CK, Surendranath V, et al. Development of human protein reference database as an initial platform for approaching systems biology in humans. *Genome Res* 2003;13:2363-71.
36. Towbin H, Staehelin T, Gordon J. Electrophoretic transfer of proteins from polyacrylamide gels to nitrocellulose sheets: procedure and some applications. *Proc Natl Acad Sci USA* 1979;76:4350-54.
37. Spring J, Beck K, Chiquet-Ehrismann R. Two contrary functions of tenascin: dissection of the active sites by recombinant tenascin fragments. *Cell* 1989;59:325-34.
38. Chiquet-Ehrismann R, Kalla P, Pearson CA, Beck K, Chiquet M. Tenascin interferes with fibronectin action. *Cell* 1988;53:383-90.
39. Fischer D, Chiquet-Ehrismann R, Bernasconi C, Chiquet M. A single heparin binding region within the fibrinogen-like domain is functional in chick tenascin-C. *J Biol Chem* 1995;270:3378-84.
40. Chiquet-Ehrismann R, Chiquet M. Tenascins: regulation and putative functions during pathological stress. *J Pathol* 2003;200:488-99.
41. Kalembey I, Yoshida T, Iriyama K, Sakakura T. Analysis of tenascin mRNA expression in the murine mammary gland from embryogenesis to carcinogenesis: an in situ hybridization study. *Int J Dev Biol* 1997;41:569-73.
42. Pearson CA, Pearson D, Shibahara S, Hofsteenge J, Chiquet-Ehrismann R. Tenascin: cDNA cloning and induction by TGF-beta. *EMBO J* 1988;7:2977-82.
43. Saito S, Yoshida M, Ichijo M, Ishizaka S, Tsujii T. Transforming growth factor-beta (TGF-beta) in human milk. *Clin Exp Immunol* 1993;94:220-4.
44. Grosvenor CE, Picciano MF, Baumrucker CR. Hormones and growth factors in milk. *Endocr Rev* 1993;14:710-28.
45. Jones PL, Boudreau N, Myers CA, Erickson HP, Bissell MJ. Tenascin C inhibits extracellular matrix-dependent gene expression in mammary epithelial cells. Localization of active regions using recombinant tenascin fragments. *J Cell Sci* 1995;108:519-27.
46. Siri A, Knauper V, Veirana N, Caocci F, Murphy G, Zardi L. Different susceptibility of small and large human tenascin-C isoforms to degradation by matrix metalloproteinases. *J Biol Chem* 1995;270:8650-4.
47. Mackie EJ. Molecules in focus: tenascin-C. *Int J Biochem Cell Biol* 1997;29:1133-7.
48. Huang W, Chiquet-Ehrismann R, Moyano JV, Garcia-Pardo A, Orend G. Interference of tenascin-C with syndecan-4 binding to fibronectin blocks cell adhesion and stimulates tumor cell proliferation. *Cancer Res* 2001;61:8586-94.

# 5. Outlook

Since the introduction of the term "degradome" various novel technological approaches emerged that have been successfully applied to decipher the substrate repertoire of a given protease on a system-wide level. The work presented herein is the first degradomic approach implemented on an astacin family member of the metzincin superfamily of metalloendopeptidases. Upon our findings hmeprin may be considered as a signalling protease mediating direct and indirect cleavage and signal transduction functions in addition to the hydrolysis of ECM proteins. Although the detailed mechanisms need to be extensively studied, hmeprin is preferentially involved in "cell growth and/or maintenance" and in "immune response". Fascinatingly, all degradomic approaches employed on metzincin metalloendopeptidases allowed the authors to come to the same conclusion: metalloendopeptidases are crucial entities in the regulation of cell homeostasis and cell environment and in innate immunity. Therefore future systems biology studies may expand our current knowledge on astacin family members and its relatives.

The highly specific cleavage of TN-C by meprinα and meprinβ at two distinct biologically important cleavage sites provided evidence that meprins are unique scissors exerting proteolytic processing. These findings comply with the current belief that metalloendopeptidases are regulatory scissors mediating proteolytic processing of target substrates rather than aberrantly destroying them. In addition, potential proteolytic processing of collagen type V by the removal of the C-propeptide and of lysyl oxidase by the removal of the N-terminal propeptide by meprin may support a previously suggested role for meprin in tissue remodelling processes. This idea is further corroborated by the fact that TN-C and the quantitatively minor collagen type V are mainly expressed during development and tissue repair. The presence of ECM proteins such as TN-C and collagen type V in human milk and culture medium demonstrated that epithelial cells, namely, mammary gland epithelial cells and MDCK cells, synthesize these ECM proteins under certain conditions. Nevetheless the main source of ECM protein production are fibroblasts derived from connective tissue. Thus the proteolytic processing of TN-C and collagen type V by meprin may in future be studied in a co-culture model using primary intestinal fibroblasts and Caco-2 cells that endogenously express meprinα. In addition, the meprin-targeted cleavage of TN-C within its cell-binding site may also be studied in a culture model using chick embryo fibroblasts. For example, chick embryo fibroblasts grown on fibronectin-coated cell culture dishes are known to spread out. In contrast, upon growth of chick embryo fibroblasts on dishes pre-coated with a mixture of intact TN-C and fibronectin in a six to one molar excess cells still attach but are not able to spread out. It will be interesting to test if TN-C fragments generated by meprin reverse this effect.

The hypothesis that meprin may be considered as a signalling protease involved in cell communication and signal transduction events needs further verification. The phosphorylation site in the C-terminal cytoplasmic tail of human meprinβ, serine 687, may mediate protein-protein interaction with diverse intracellular binding partners. Therefore upon phosphorylation of serine 687 in meprinβ different intracellular signalling events may be triggered. A first step towards identifying putative binding partners may be the generation of meprinβ mutants by site-directed mutagenesis where the serine 687 is replaced by an alanine. Histidine-tagged wild-type and mutant meprinβ forms may then be overexpressed in stably transfected MDCK cells treated with PMA to induce phosphorylation. The histidine-tagged wild-type and mutant meprinβ forms may then be affinity-purified using a nickel column. Cognate binding partners may be found in the column eluate by comparing wild-type versus mutant meprinβ. The proteins of these eluates may then be separated by 2-DE and subsequently identified by means of LC-MS/MS.

# 6. Acknowledgements

I wish to acknowledge and thank my supervisor Prof. Erwin E. Sterchi and my tutor Prof. Dr. Ulrich Baumann for having given me the chance to work on a proteomic topic during my PhD thesis; my external examiner Prof. Dr. Robert Beynon, Dr. Duncan Robertson and Lynn McLean for the excellent time in Liverpool during summer 2003 and for having taught me in proteomics and mass spectrometry; Dr. Manfred Heller and Daniel Stalder for identification of proteins by mass spectrometry; the members of the Sterchi group, Ursula Luginbühl, Beatrice Oneda, Maya Huguenin, Sandra Rösmann, Mario Noti, Benedikt Gehr and Dr. Daniel Lottaz for the nice working environment; Prof. Dr. Matthias Chiquet for the generous gift of chick tenascin-C and domain-specific monoclonal antibodies; Dr. Eduard B. Babiychuk for the generous donation of annexin A1 and vinculin antibodies; Dr, Christoph Becker-Pauly and Prof. Dr. Walter Stöcker for the generous endowment of recombinant human meprin$\alpha$ and $\beta$; Prof. Dr. Bernhard Erni for free access to Fuji Film Fluorescent Image Analyzer FLA-3,000R and AIDA software and finally, all my friends and people that I did not mention by name.

Mein besonderer Dank geht an meinen Vater German, der trotz schwerer Krankheit mich stets unterstützt hat; an meine starke Mutter Margrit, die unsere Familie während schwierigen Zeiten zusammengehalten hat und an meinen Bruder André, der immer an mich geglaubt hat. Dank eurer Stärke und Liebe habe ich meine Arbeit zu Ende gebracht.

This work was funded by the Swiss National Science Foundation (SNSF) (grant 3100A0-100772 to E.E.S.) and the European Science Foundation (ESF) Integrated Approaches for Functional Genomics (grant 0341 to D. A.).

Die VDM Verlagsservicegesellschaft sucht für wissenschaftliche Verlage abgeschlossene und herausragende

# Dissertationen, Habilitationen, Diplomarbeiten, Master Theses, Magisterarbeiten usw.

für die kostenlose Publikation als Fachbuch.

Sie verfügen über eine Arbeit, die hohen inhaltlichen und formalen Ansprüchen genügt, und haben Interesse an einer honorarvergüteten Publikation?

Dann senden Sie bitte erste Informationen über sich und Ihre Arbeit per Email an *info@vdm-vsg.de*.

**Sie erhalten kurzfristig unser Feedback!**

VDM Verlagsservicegesellschaft mbH
Dudweiler Landstr. 99         Telefon  +49 681 3720 174
D - 66123 Saarbrücken         Fax      +49 681 3720 1749
**www.vdm-vsg.de**

Die VDM Verlagsservicegesellschaft mbH vertritt

Printed by Books on Demand GmbH, Norderstedt / Germany